THE INTERNET
OF THINGS

SAMUEL GREENGARD

D0180123

The MIT Press | Cambridge, Massachusetts | London, England

© 2015 Massachusetts Institute of Technology

MIT Press books may be purchased at special quantity discounts for business or sales promotional use. For information, please email special_sales@mitpress.mit.edu.

This book was set in Chaparral by the MIT Press. Printed and bound in the United States of America.

Library of Congress Cataloging-in-Publication Data

Greengard, Samuel.
The internet of things / Samuel Greengard.
    pages   cm.—(MIT press essential knowledge series)
Includes bibliographical references and index.
ISBN 978-0-262-52773-6 (pbk. : alk. paper)
1. Embedded computer systems—Popular works. 2. Internet of things—Popular works. I. Title.
TK7895.E43G74 2015
384.3—dc23

2014042259

10   9   8   7   6   5   4   3   2   1

CONTENTS

SERIES FOREWORD

The MIT Press Essential Knowledge series offers accessible, concise, beautifully produced pocket-size books on topics of current interest. Written by leading thinkers, the books in this series deliver expert overviews of subjects that range from the cultural and the historical to the scientific and the technical.

In today's era of instant information gratification, we have ready access to opinions, rationalizations, and superficial descriptions. Much harder to come by is the foundational knowledge that informs a principled understanding of the world. Essential Knowledge books fill that need. Synthesizing specialized subject matter for nonspecialists and engaging critical topics through fundamentals, each of these compact volumes offers readers a point of access to complex ideas.

Bruce Tidor
Professor of Biological Engineering and Computer Science
Massachusetts Institute of Technology

ACKNOWLEDGMENTS

Writing a book is a huge investment in time and energy. This one was no exception. I'd like to thank Eileen Feretic at *Baseline* magazine for assigning a magazine story that served as the seed for many of the concepts and ideas that appear in this book. I'm also grateful for the insights and direction provided by Andrea Linne, editor of *RFID Journal*. She took time from her busy schedule to provide some background and history about the Internet of Things and connected devices. Of course, the IoT is an incredibly complex topic with many nuances. I appreciate those who allowed me to interview them and pluck their knowledge—some of which appears in this book.

A big thank-you also goes out to Marc Lowenthal, my editor at MIT Press, who provided clear direction and was incredibly easy to work with from beginning to end. Ditto to two anonymous reviewers who provided some excellent suggestions on how to improve the manuscript. And thanks to Dana Andrus, an MIT Press senior editor who read the manuscript and caught all the tiny but important errors. Finally, I'd like to tip my cap to my partner, Patricia Hampel Valles, who dutifully proofread the manuscript, spotted a number of errors and cognitive glitches, and, most important, spent quite a few evenings and weekends going solo while I researched and wrote the book. Also a big hug to my sons Evan and Alec Greengard, who light up my life every single day.

INTRODUCTION

It's incredibly easy to overlook the full impact of technology on our world. The wheel made it possible to move and transport things and people. It changed everything from agriculture to political governance. The lightbulb illuminated homes and businesses—eventually changing the way architects designed structures and entire cities were laid out. The automobile introduced fast point-to-point travel for individuals—redefining the way we live and work. And the computer introduced a digital world with data that could be stored and shared in new and remarkable ways. This ushered in massive changes in how people act ... and interact.

Each of these inventions, and countless others—from refrigerators and sewing machines to telephones, typewriters, and cameras—eventually settled into the mainstream of society and unleashed massive political, social, and practical changes. They became things that people use in their daily life—and to a large extent take for granted. Yet they also rewired the way people approach countless tasks and redefined how we interact, communicate, and go about our daily work.

In 1957 Joe M. Bohlen, George M. Beal, and Everett M. Rogers at Iowa State University introduced the now commonly accepted notion of a technology adoption curve.

Together, they forwarded the concept that any new product or solution follows a fairly predictable trajectory that roughly approximates a bell curve. The earliest adopters are referred to as *innovators*; the second phase is named *early adopters*; then come the *masses* followed by the *laggards*. This model still holds true, though over the last couple of decades the life cycle has accelerated at warp speed. In some cases the model now compresses years or decades into months.

The Internet of Things lies at the epicenter of this shock wave—and it is just getting started. While it will someday serve as the practical framework for life and business, it now resides somewhere between the innovator and early adopter phases. Connected devices have existed in one form or another since the introduction of the first computer networks and consumer electronics. However, it wasn't until the Internet emerged that the idea of a globally connected planet began to take shape in earnest. In the 1990s researchers theorized how human and machine could weave together a completely new form of communication and interaction via machines. That reality is now unfolding before our eyes.

Although it's impossible to identify a single happening that sparked this revolution, it's safe to say that Apple's introduction of the iPhone in 2007 was a crystalizing event. It put smartphones in the hands of the masses. It put real-time point-to-point communication on the map with a powerful device that could be held in a hand. Consider: In January 2008, Apple had sold approximately 3.7 million units. By June 2014, the number had topped 500 million. Today, the total number of smartphones in use worldwide is somewhere in the vicinity of 1.9 billion. By 2019, Swedish telecom firm Ericsson estimates the number will exceed 5.6 billion.

Each of these phones has an array of chips that can record data, voice, video, audio, motion, location, and much more. Moreover these phones allow users to connect to other machines. Smartphones can serve as remote controls, dashboards that display personal data, and information feeds. They can receive alerts and notifications for events, and they can hold boarding passes, electronic tickets, payment systems, and much more. Together, they can also tap into social media data and crowdsourcing to create news ways of collecting, managing, and analyzing events in the physical world.

At the same time, RFID technology has matured, sensor technology has leapt forward, miniaturization has accelerated, and computer software has taken a giant leap forward. The convergence of these technologies—along with nearly ubiquitous wireless networks and cloud computing—has introduced the concept of robotic insects and animals, nanobots and microbots that can exist inside humans, and drone fleets that can accomplish tasks in the sky above. Make no mistake, we are entering a brave new world of immersive and embedded technology. It's a world that, at first glance, may seem much more like science fiction than science fact.

But it is fact. The Internet of Things offers both a telescope and a microscope into the once invisible world between people, machines, and physical objects. By tagging objects and imbuing them with Internet connectivity it's suddenly possible to not only track the objects and collect new types of data but also combine these data to generate a greater level of information and knowledge. It boldly goes where no data scientist could have ventured—let alone imagined—only a few years ago.

It's as if the rules of earthly physics have been rewritten on the fly. The Internet of Things connects human and machine intelligence in new, entirely remarkable, and sometimes scary ways. It can make sense of the motion between and among things, including people, animals, vehicles, air currents, viruses, and much more. It can recognize relationships and predict patterns that are far too complex for the human mind and senses to grasp—such as the condition of a bridge or a roadway or the physics of the atmosphere on a block-by-block scale. The IoT can also support systems that operate independent of human oversight and, incredibly, get smarter on their own over time by adapting an underlying algorithm.

The IoT is the second wave of a powerful digital revolution that began with the widespread adoption of computers in the 1970s and 1980s. And, like all revolutions, it promises to create scores of winners and losers. The Internet of Things will introduce new products and services and make many existing offerings completely obsolete. The technology will eliminate jobs but introduce new lines of work. Connected systems will ripple through education, government, and business and fundamentally remap and rewire actions, behavior, and social norms. The technology will affect everything from the way people vote to the way we eat at restaurants and take vacations.

However, the potential gains won't come without a good deal of pain—and plenty of unintended consequences. The future could introduce new types of crime, weapons, and warfare. It could also create significant political and social problems by, among other things, contributing to a growing disconnect between people. It will certainly cause society to more closely examine the notion of privacy and security.

While it is impossible to know where exactly the Internet of Things will take us, it's entirely clear that a more technology-centric world is in the cards. We will live in automated homes, drive smart vehicles on networked roads, shop in highly interactive stores, and connect to medical and wellness products that redefine our basic approach to health. Within a decade, we will use a mind-bending array of other smart systems in our daily life.

This book provides a guided tour through the emerging Internet of Things. Think of it as a Digital Carousel of Progress. We will examine the origins of the IoT in chapter 1. In the beginning, there were personal computers and the Internet. They ushered in global communications on a person-to-person level. The Internet serves as the electrical wiring for the IoT. It makes real-time communication and data sharing possible on a mass scale.

Chapter 2 examines the impact of mobility and cloud computing—and how these two powerful technologies create a conceptual and practical framework to support a connected world. This includes communications but also apps and embedded capabilities that make it possible to build out an infrastructure that supports tools such as social media and big data. Together, these technologies unlock the greater value of the IoT.

Chapter 3 will peer into the Industrial Internet and machine to machine communication (M2M)—the basis for smart manufacturing, end to end supply chain visibility, better public safety, and much more. Huge efficiencies of scale result from the IoT. There is also the potential for significant costs savings through greater automation and sensor-based analytics.

In chapter 4, we take a look at the growing array of smart consumer devices and services that redefine the way we interact with the world. This includes everything from Fitbit fitness wristbands to smartphone enabled front door locks and lighting systems. We'll examine how the concept of connected devices has evolved and matured and where it is heading in the months and years ahead.

Chapter 5 dives into the practical and technical challenges of building the IoT, including the development and integration of more advanced hardware, software, and sensors. We will also assess the need for technical and industry standards and, in the end, understand what it takes to put all the data to use effectively.

Chapter 6 offers insights into the concerns, risks, and problems associated with a connected world. Already, serious concerns exist about whether this technology will dumb down society, lead to greater inequality, and expand the digital divide. But it raises other questions as well: *Could automation cause massive unemployment and downward mobility. Could it cause more crime or new types of terrorism and warfare? How might it change the legal system? What about the growing problem with digital distraction?* No less important: *How do we approach security and privacy in an era where almost no movement or activity goes unnoticed or unrecorded?*

Finally, in chapter 7, we'll speculate about how the future might unfold and how the Internet of Things will impact society on a long-term basis. We will hear what different experts have to say about the IoT and view possible scenarios for life and work in 2025.

The Internet of Things will touch almost every part of our lives in the years ahead. While it's impossible to address every aspect of the topic in this book, the pages ahead offer a glimpse into a world that promises to alter our lives faster and more profoundly than any technology in history. It's not a question of whether the IoT will take place, it's a matter of how exactly it will happen and how much it will change the world.

THE INTERNET CHANGES EVERYTHING

A Day in the Life

It's Monday morning at 7 am. My Fitbit Force wristband vibrates to wake me up. After a few minutes, I reach over and pick up my iPhone so I can check for e-mail and other messages. I tap the icon for the Fitbit app and glance at my sleep pattern during the night, including how long it took to fall asleep and how many times I woke up. I step out of bed and shuffle to the bathroom, where I weigh myself using a Fitbit scale that automatically sends the data to a server in the cloud. It in turn crunches the numbers and presents feedback via a website or a smartphone app. I can track my weight, body fat, food intake, water consumption, and overall activity level.

At breakfast, I use an app on my iPhone, *MyFitness-Pal*, to scan the barcode on the oatmeal package. It taps a database on the Internet with more than 3 million entries and provides calorie and nutrition data. Afterward, I drive to the gym to work out. I enter an ID on the treadmill and it tracks my activity, including how far I've traveled, how much I've climbed and how many calories I've burned. After concluding my run on a treadmill, the machine sends my workout data to MyFitnessPal, which in turn connects to the Fitbit app on my phone. The combination of these devices and apps delivers a fairly complete picture of my daily activity and food consumption. I can tell if my calorie and exercise levels are on target or lagging. I can peer into my nutrition data and see if I'm well hydrated through charts, graphics, and dashboards.

When I step back into my house, I glance at my driving data on an iPhone app, Metromile, which uses a device in my car to gauge my mileage, fuel costs and more. After I take a shower, I pick up my iPad, check Facebook and sort through a stack of e-mail messages. Next, I step into my home office, where I begin working at my desktop computer. A bit later in the day, I remember that I'll be out of town during the weekend. So I program my Ecobee Internet-enabled thermostat for vacation mode from my phone. I also remember to program a temporary lock code for the Kevo lock on the front door, so that a neighbor can get inside to water the plants.

After completing work, I prepare dinner and then switch on Netflix using a software-based Harmony remote on the phone. I watch the film from a Blu-ray DVD player that connects to the Internet via my wireless network. At dusk, a WeMo light switch automatically switches on the porch light, based on daily sunset information for my geographic location. It grabs the data every day and always stays current on changes. A few minutes later, I receive an alert on the phone. It notifies me that my garage door has been open for 30 minutes. It turns out that one of my kids has left the door open after taking out the trash. I push a button in the app and close the garage door.

At 11:30 pm, WeMo switches off my front porch light. I climb into bed and begin reading an article in a print magazine. I decide I want to clip it electronically. I grab my phone and pull up an app called DocScanner. It lets me input the article into Evernote, which syncs through the cloud so that it can be viewed again in a project folder on any of my devices. I set my Fitbit alarm to wake me up the next morning, switch off the light and drift off to sleep.

This scenario isn't fiction. It represents a realistic snapshot of a typical day in my home, which hardly qualifies as a state-of-the-art lab for connected devices. In fact my router now displays a total of 19 wireless clients—each with an IP address—including computing devices, media players, home automation gear, and more. Many of these devices rely on mobile apps and all connect to the Internet of Things. For better or worse, these connected devices eliminate manual tasks and deliver entirely new ways to access digital content; gain insights; and manage locks, doors, lights, and thermostats. Some of these connected devices also deliver the benefit of energy savings by helping devices operate smarter and more efficiently. Some others provide better security.

A Brief Look at How We Got Here

It's incredibly easy to overlook just how profoundly the world has changed over the last couple of decades. Not long ago, before the Internet, mobile devices, and cloud applications existed, data resided mostly on gigantic mainframes and, later, on the hard drives of personal computers. Most of these machines were little more than stand-alone islands in a vast sea of computers. Getting data from one device to another was no simple task. Other than a fortunate few who had access to a local area network (LAN), floppy drives were the order of the day.

By today's standards, the process of transferring data from one floppy drive to another was slow and cumbersome. These drives also provided extremely limited capabilities. For one thing, the drives were bulky. The first storage disks measured 8 inches in diameter. What's more, the media format stored only about 80 kilobytes of data. That's about 40 pages of plain text. In the mid-1970s, the introduction of the 5 1/4-inch floppy disks pushed media capacity to 110 kilobytes and by 1982, 1.2 megabytes. By the late 1980s, 3 1/2-inch disks held about 2.4 megabytes of data. While this represented a significant advance at the time, it is almost unfathomable by today's technology.

Physically transporting data on magnetic media was equally challenging. A person had to mail or carry disks to another location. This meant that it could take hours or days to transport any significant volume of data. In the 1980s and even the 1990s, it wasn't unusual to install a software program using anywhere from 10 to 20 floppies. The process could take more than an hour and tie up a computer during that time. During this era, PCs did not have today's multiprocessors and multitasking capabilities. Although manufacturers introduced alternative media that pushed up data capacity, including once popular Zip Drives, the gains mostly revolved around greater convenience in managing data rather than connecting systems more efficiently.

The widespread adoption of computer networks in the 1990s changed all of this. Ethernet and LANs allowed organizations to share data internally—and sometimes with business partners and others outside the four walls of the enterprise. However, the expensive and proprietary nature of these networks coupled with relatively slow transfer speeds limited their value—and overall adoption rates. Connectivity and connectedness were still out of reach for the vast majority of people—and machines.

Oftentimes, remote users were forced to dial into a computer—usually a mainframe—via a 300 bps modem in order to send or receive a file. Setting up protocols and transferring data was frequently a daunting task. It could take minutes or longer to send a short text file and any type of large file could devour system resources and render the computer essentially useless for hours. By today's standards, it was the wild frontier of data transfer. It was the digital equivalent of Magellan attempting to circumvent the globe using a wooden carrack ship with sails that relied on the wind.

Then, in 1995, after years of discussion, the Internet and World Wide Web were commercialized. Born out of research on packet-based networks during the 1950s, the original ARPAnet (Advanced Research Projects Agency Network) had evolved from its humble introduction in 1969 into a far more robust Internet Protocol (IP) network (IP, along with Transmission Control Protocol, or TCP, serves as the protocol for establishing a virtual connection between devices or systems). An ongoing series of technical advances—as well as massive leaps in computing power—led a legion of private groups to push for an open Internet. After the US government decommissioned the network, then known as the *National Science Foundation Network*, a new era was born. The framework for global connectedness had been established.

The first Internet connections took place mostly using dial-up modems and a web browser called Mosaic, which was introduced by Marc Andreesen (the company later became Netscape). The Mosaic browser was based on the earlier work of Tim Berners-Lee, an Oxford University graduate then working at CERN, the European Particle Physics Laboratory. He invented the first web browser, World Wide Web (later renamed Nexus), in 1990 using a then-powerful NeXT computer.

At first, connection speeds were painfully slow. It could take minutes for large pages to load on the Web and a user was typically connected only while logged in through services such as America Online (AOL), CompuServe, and EarthLink. Except for a few large universities, research facilities, businesses, and government institutions, broadband wasn't part of the online picture and wouldn't be for another several years. In 2000, about 3 percent of the US population had a broadband connection at home. By August 2013, the figure had swelled to about 70 percent of US households.[1] The numbers are even higher in a handful of countries.

Like the first railroad tracks laid down during the Industrial Revolution, the framework for a wired and connected future suddenly existed. The inventors of the Internet—including Robert E. Kahn and Vint Cerf—envisioned a world where networks connected to other networks—thus creating an interconnected fabric of networked systems. They

The inventors of the Internet—including Robert E. Kahn and Vint Cerf—envisioned a world where networks connected to other networks—thus creating an interconnected fabric of networked systems.

foresaw a world with smarter machines that would spawn remarkable capabilities and incredible transformation. In a 1999 interview I conducted with Cerf for *America West* magazine (the in-flight publication for now defunct America West Airlines), he elaborated on his goals at the time:

> The principle goal, back in 1973, was to create a way for computers to communicate with one another. Back then, we had developed different computer networks that functioned independently. It was very clear that all these systems were of limited value unless they could share information through a common language. We certainly didn't want to wind up with a situation parallel to the 1910s and 1920s, when a business had a dozen different telephones sitting on a desk—all using a different proprietary system and requiring a person to know which telephone service to use to reach someone else. So, we invented a protocol called TCP/IP, which allows computers and various networks to interconnect.
>
> We knew the technology was powerful, and we knew that it offered enormous possibilities. But at the time, we were working on computers that were million dollar behemoths. They filled entire rooms. They weren't units that you stuck in your briefcase and carried home. Perhaps even more interesting, though, is that it's difficult to envision what happens

when extremely large numbers of people gain access to a technology, such as the Internet. It's a bit like being the inventor of the automobile and imagining a few dozen of them, not knowing how 50 million or 100 million of them would affect the attitudes, customs, behavior and actions of the entire country … and the world.

To be sure, high-speed Internet access is now ubiquitous in developed nations. Moreover, with the introduction of mobile devices and mobile broadband over cellular networks, an always-on, always-connected culture has emerged. Credit the 2007 introduction of the iPhone and the 2010 introduction of the iPad with changing the stakes and lighting the fire for today's emerging IoT. Although a series of manufacturers had previously introduced so-called smartphones and PDAs that could connect to the Internet, these devices were clunky, slow, and provided extremely limited features and capabilities. Many such devices simply synced calendars, contacts, and basic data. And most did an abysmally bad job at doing much of anything well—other than a voice call.

But the foundation for a connected world—and connected devices—was born. Today every connected device receives an IP address, and every address allows every device to connect to other devices, including smartphones, tablet computers, gaming consoles, automobiles, refrigerators, washing machines, lighting systems, front door locks, electronic toll devices in vehicles, and much more. Over the last several years a variety of systems and platforms have emerged based on IP. In fact IP is now the standard conduit for communication, entertainment, shopping, business transactions, and an array of other tasks and activities.

Driving this trend is the growing digital nature of machines and systems. Only a couple of decades ago, audio and video recorders used tape, cameras snapped pictures with film, remote controls were built with hardware and music played from records, tapes, or compact discs. People usually printed pieces of paper and sent the paper to others through the mail or via a fax machine. In this world of both analog devices and digital devices, each machine delivered a separate function, and there was often no way to move data between devices—except with physical media. That, of course, limited usability and convenience.

But today a remarkable array of functions and features are packed inside a typical computing device, such as a tablet or smartphone. Increasingly, thanks to the *lingua franca* of binary code and Internet Protocol, these devices consolidate the functionality of several devices in the past and operating them is a simple and straightforward proposition. Commands, functions, and coding that would have once required an army of developers with intimate knowledge of a programming language now take place at the tap of a finger or the utterance of a spoken word. Indeed a user requires virtually no knowledge of computing device to use one and accomplish an array of seemingly complex tasks.

The net effect? Digital technology is crumpling entire industries and driving radical changes in many others. Conventional cameras and film have largely disappeared, stand-alone video and audio recording devices are disappearing, paper maps are vanishing, landline phones are on the way to going extinct, and traditional books and magazines are becoming a chapter in history. What's more, even singular and dedicated devices are increasingly connected to the Internet. DVD players stream content from remote servers, navigation systems on automobiles display traffic based on sensor and satellite data, and bathroom scales upload information to the Internet. A growing array of industrial machines—from medical equipment and farm equipment—also transmit data to the Internet, where all these data are slotted into databases, combined with other data, and analyzed.

All these digital devices add value to products and services. Suddenly a $75 handset becomes a $600 smartphone that redefines our world. According to Cisco Systems, which has established an Internet of Everything (IoE) Index, businesses now generate $613 billion of additional profits annually as a result of connected devices, but this represents only about 50 percent of the potential of the Internet of Things. The figure could reach $14.4 trillion in net profits within a decade, Cisco estimates.

A New Deal

To be sure, we now live in a connected world. Marshall McLuhan's Global Village has arrived and the digital age is flourishing. Today there are an estimated 7 billion Internet users worldwide. Cisco Systems estimates that approximately 12.1 billion Internet connected devices were in use in April 2014, and the figure is expected to zoom to above 50 billion by 2020. In fact the networking firm says that about 100 "things" currently connect to the Internet every second but the number will reach 250 per second by 2020.[2] Overall, the Internet Business Solutions Group at Cisco Systems estimates that more than 1.5 trillion "things" exist in the physical world and 99 percent of physical objects will eventually become part of a network. Of course, time will tell whether this estimate is overly optimistic or grounded in reality.

In the meantime these things are taking new shapes and forms. It's no longer only computers and smartphones that connect to the Internet. The list includes parking meters, thermostats, health monitors, fitness devices, traffic cameras, tires, roads, locks, supermarket shelves, environmental sensors, even livestock and trees. What's more, these capabilities are growing exponentially as a mélange of digital technologies intersect, prices for hardware and software continue to drop, persistent connectivity becomes faster and more reliable, and developers learn to better integrate devices, apps, platforms, and more.

Alone and together, these devices provide new features and entirely new capabilities for businesses and consumers. For instance, it's possible to adjust a thermostat, switch lights on and off, and provide temporary access codes for the front door lock—via a smartphone located across town or on the other side of the world. In addition, the technology introduces opportunities to put data to work in new and intriguing ways using social media, crowdsourcing, geolocation data and, ultimately, big data and analytics. The latter incorporates today's large and growing datasets. Some observers believe that all these data will soon serve as a genuine currency that will impact businesses, stock valuations, and merger and acquisition activity.

The Internet of Things makes it possible for epidemiologists to track the spread of viruses in near real time. A grocery store can analyze how people shop and the products they view and buy as an individual walks through the store. A clothing manufacturer can view changing fashion tastes and trends as they happen. A pharmaceutical firm can understand consumption patterns in real time. And a city can crunch data from sensors and other systems to better manage congestion, waste management, utilities, natural resources, and much more. No industry will be left untouched by the Internet of Things. The technology brings intelligence and a far greater level of insight and understanding to a vast array of physical and virtual systems.

Defining the Terms, Understanding the Concept

By now it should be clear that the Internet of Things quite literally means "things" or "objects" that connect to the Internet—and each other. This could be almost anything—a computer, tablet or smartphone, fitness device, lightbulb, door lock, book, airplane engine, shoes or football helmet, to name a few. Each of these devices or things has a unique identification number (UID) and an Internet Protocol (IP) address. These objects connect via cords, wires and wireless technology, including satellites, cellular networks, Wi-Fi, and Bluetooth. They use built-in electronic circuitry as well as radio frequency identification (RFID) or near-field communications (NFC) capabilities that are added later via chips and tags. Regardless of the exact approach, the IoT involves the movement of data to enable processes from across the room or somewhere on the other side of the world.

But, within the vast category of the Internet of Things, there are a few key distinctions and nuances. At this point, some basic definitions are in order. First of all, the term "connected devices" refers to devices that exchange data through the standard Internet and gain some benefit by connecting over a network—sometimes a private or closed network. Connected devices don't necessarily have to connect to the Internet of Things, but they increasingly do.

What's more, they extend connectedness far beyond computers and into all the nooks and crannies of the world.

Two basic types of connected objects exist: physical-first and digital-first, according to a May 2014 briefing paper from ABI Research, *Internet of Things vs. Internet of Everything: What's the Difference?* The former consist of objects and processes that do not typically generate or communicate digital data unless augmented or manipulated "whereas the ones belonging to the digital-first domain are capable of generating data, and communicating it on for further use, inherently and by design."[3]

It's an important distinction. For instance, a hardcopy of a book or a vinyl record are examples of a physical-first object. However, an e-book and an MP3 audio file are digital-first. They are essentially native to the digital world because they are comprised of binary code rather than physical substance. Likewise a brick-and-mortar store is physical-first and an online store is digital-first. While many physical-first objects can be tagged using digital tools and technologies, such as RFID, they typically do not provide the same level of data and a similar level of insight. For instance, a marketer can track the way a reader uses and reads an e-book by studying every click and tap. A tagged hardbound book may reveal its location, but it will provide little more data because paper and ink aren't digital. However, depending on the task, say a librarian looking for a misplaced book, this may still prove valuable.

A key tool that makes it possible to bring physical devices into the digital realm is RFID. The technology relies on microchips that pull data from sensors built into the machines or chips that reside on or in a device. RFID uses both "active tags" with a power source—often a battery—and "passive tags" that do not require a battery or other power source. Both allow nearby RFID readers to collect and exchange data with computers. When an RFID chip is within range of the reader, it automatically sends a signal and data to a computer.

Passive RFID is particularly compelling because it doesn't require a power source, the tags can function for twenty years or longer, and they cost only a few cents each. A passive tag pulls the necessary power from a nearby reader. A coiled antenna in the device creates a circuit and the tag forms a magnetic field.

Another term that enters the IoT sphere is "Industrial Internet," which revolves around machines equipped with sensors, thus making them "smart." These devices often serve as the plumbing or IT foundation for the IoT. For example, industrial machinery or a delivery truck may stream data to the IoT. These data could also be combined with other data to further enhance the features and overall value. Within the Industrial Internet, communication typically takes place in three different ways: machine to machine (M2M), human to machine (H2M), and machine to smartphone (M2S) or other device, such as a tablet.

Obviously each has different implications and ramifications, which we'll examine throughout the book.

What makes the Internet of Things so powerful is that it connects physical-first products and items to each other as well as connecting them to digital-first devices, including computers and software applications. This makes it possible for all these devices to interact on a group or multipoint basis and share data in real time—often through cloud computing. Moreover, when all these machines connect to people using various computing devices—essentially the Internet of Humans (IoH)—an entirely new conceptual framework is born.

The sum of all this is the Internet of Everything (IoE), a term created by networking firm Cisco Systems. It represents a more evolved and advanced state where physical and digital worlds are blended into a single space. The value of this human–machine world is unlocked as more and more capabilities become interconnected and intertwined. As ABI notes: "As the controlled Things become smarter, enabled by machine learning and artificial intelligence, it is likely that the need for human input will gradually decrease. The Things that today require such guidance to be aware of what the human user prefers will in the future be immersed into the environment in which they operate. In this sense, it can be argued that the IoH is a stepping stone towards a more immersive level of intelligence" (p. 5).

The concept of connected devices first emerged in the early 1990s. At that time researchers at the Auto-ID Center at MIT began pondering the idea of building a system that would allow devices in the physical world to connect via sensors and wireless signals. In 1999, Kevin Ashton, cofounder of the Auto-ID Center at MIT (it was shut down in 2003 and EPCGlobal was then launched to commercialize so-called electronic product code, or EPC, technology), coined the term "Internet of Things." As early as 1997, he had pondered the possibility of using RFID to help manage the supply chain at consumer products firm P&G, where he had worked as an assistant brand manager. When the center swung open its doors two years later, Ashton played a pivotal role in establishing a global standard for RFID, before becoming a high-tech entrepreneur and creating his own startup companies.

At the time, researchers viewed RFID as a necessary precursor to the Internet of Things. The technology—along with near-field communication, barcodes, QR codes, and digital watermarking offered a way to bridge the gap between physical objects and the virtual world. As Ashton pointed out in a 2009 *RFID Journal* article, the Internet of Things or IoT, alters the equation from human-based data input to both human- and machine-based data input. While most data on the Internet currently takes the form of text files, messages, audio, photographs, and video files, the IoT grabs new and different data, it combines data in different ways and it allows humans and machines to gain broader and deeper insights.

The article, titled *That 'Internet of Things' Thing*, included Ashton's thoughts:

> Today computers—and, therefore, the Internet—
> are almost wholly dependent on human beings
> for information. Nearly all ... data available on the
> Internet were first captured and created by human
> beings—by typing, pressing a record button, taking
> a digital picture or scanning a bar code. Conventional
> diagrams of the Internet include servers and routers
> and so on, but they leave out the most numerous
> and important routers of all: people. The problem is,
> people have limited time, attention and accuracy—all
> of which means they are not very good at capturing
> data about things in the real world.
>
> And that's a big deal. We're physical, and so is
> our environment. Our economy, society and survival
> aren't based on ideas or information—they're based
> on things. You can't eat bits, burn them to stay warm
> or put them in your gas tank. Ideas and information
> are important, but things matter much more. Yet
> today's information technology is so dependent on
> data originated by people that our computers know
> more about ideas than things.
>
> ... We need to empower computers with their own
> means of gathering information, so they can see, hear
> and smell the world for themselves....[4]

Make no mistake, the potential for a connected world is enormous. The IoT can dive into the nooks, crannies, gaps, and wormholes that exist in an imperceptible and often invisible world that extends far beyond human eyes, ears, smell, and consciousness. It creates new types of networks and systems—and entirely different pathways for data, information and knowledge to travel. Along the way, and with the right input and analysis, computers and humans can decipher the codes that dictate the physics of the planet and various events. This might translate into something as straightforward as knowing when a package of food has expired or a machine is about to fail or something as complex as managing smart cars in an automated grid that spans a city. Or, using smart shelves and tagged products that can identify a person, know his or her preferences and dispense relevant information and coupons at the right place and at the right time.

There are also IoT-related concepts that, although bordering on science fiction, will likely appear in the real world over the next couple of decades. For example, RFID sensors and other devices implanted in the human body or worn on the body could gather data and use the IoT to transmit specific information about blood pressure, blood sugar, heartbeat, and other vitals while also monitoring medication dosage. This could ensure that the right amount of medicine is dispensed at all times. So-called nanobots could also help physicians monitor elderly patients and

The IoT can dive into the nooks, crannies, gaps and wormholes that exist in an imperceptible and often invisible world that extends far beyond human eyes, ears, smell, and consciousness. It creates new types of networks and systems—and entirely different pathways for data, information, and knowledge to travel.

know when there's reason for concern or when something is wrong. If a monitor detects a problem, it could alert a physician or emergency responder immediately. Likewise 3D printers could fabricate an array of objects on demand, including replacement organs for the human body.

The ideas are limited only by human imagination and creativity. Because the IoT is still in its infancy, there are many technical and engineering issues to tackle, including developing better and longer lasting batteries for mobile devices, building smaller devices and packing more sensors into existing smartphones and other devices, identifying ways to embed sensors in everything from clothing to machinery, perfecting miniaturization, developing better algorithms to sort through all the data and keep the signal-to-noise ratio low, and developing standards and platforms that enable data sharing and widespread compatibility. Today any sensor, physical or virtual, can be transformed into a source of data. And all these data, once collected, can be analyzed, so the opportunities are infinite.

A Sense of Security

While the Internet has ushered in a world where news, information, transactions, and social interactions take place at the speed of light—and in many cases produce enormous time gains and cost savings—it's also obvious that it has become a hotbed for hacking, data breaches, malware, cyber attacks, snooping, and myriad other problems. These breaches already place personal and financial data from millions of consumers at risk—and fuel an epidemic of identity theft and fraud. The Identity Theft Resource Center estimated that 614 major known breaches occurred in 2013—a 30 percent increase over the previous year.[5] They involved more than 92 million records in the United States alone. These breaches touched retail, health care, financial services, and every other sector. Worse, there's no sign that the problem is slowing.

The Internet of Things ratchets up the risk—and it also creates entirely new threats. We'll take a closer look at security and privacy risks in chapter 6, but it's clear that many observers see a somewhat dystopian future ahead. According to a 2013 survey conducted by ISACA (formerly the Information Systems Audit and Control Association), 92 percent of the public expresses concerns about the information collected by Internet-connected devices. Already hackers have broken into an array of IoT devices, including automobiles, video cameras, and baby monitors. Additionally so-called white hat hackers (so named because they identify vulnerabilities in order to improve software rather than exploiting it for any gain) have discovered vulnerabilities in connected medical equipment, including insulin pumps, ventilators, and defibrillators.

While a hacker commandeering a video camera may evoke feelings of fear and creepiness, programming the brakes to fail on a vehicle or a pacemaker to stop functioning on a person could have dire consequences. It's clear that engineers, designers, developers, and security experts must sort through an array of issues before IoT devices and connectivity can march forward. The concern is particularly acute in areas such as industrial machinery, health care, and transportation—where people's lives, health, and welfare are at stake.

Putting the Pieces Together

The Internet is continuing to evolve. Ongoing advances in semiconductors, microelectronics, computer design, storage devices, cloud architecture, and more, are ushering in new capabilities and features. Higher bandwidth cellular data networks and faster Wi-Fi are introducing a more robust infrastructure to support the IoT. A report produced by Cisco Systems and consultancy Global Business Network (GBN), *The Evolving Internet*,[6] points out that the original architects of ARPAnet never anticipated the Internet of today—including the sheer number of users and the security risks it presents. Although the Internet is comprised of a mesh of digital impressions, storage systems, fiber, radio frequencies, transmissions, switches, screens, and terminals, it also involves a complex array of relationships across technologies, applications, players, and policies, the report notes.

All these factors will play a role in shaping the future of the Internet and determining the direction of the IoT. Ultimately one thing is clear: the breadth and depth of the Internet is expanding beyond human input and interaction. IP-based digital technology intersects with our lives every day, and increasingly, cloud computing, mobile apps, crowdsourcing, social media, and big data play a crucial role in influencing behavior and defining human interactions. As the Internet of Things takes hold and burrows deeper into our lives, it promises to further redefine everything from health care and retailing to entertainment and politics. We will wear clothes that connect to the Internet and drive cars that are smart enough to take the fastest route on any given day.

The Internet of Things is nothing less than a disruptive event. Some have referred to it as the Industrial Revolution 2.0. Others have described it as an upheaval that will change the world more than any technology platform that has come before it. Although it's impossible to predict the specific course and the eventual outcome, it's apparent that the line between human and machine will continue to blur. What's more, networked and independently functioning machines will increasingly assume their own intelligence. In the time it took you to read the last paragraph, another few thousand devices have come online.

The future has arrived.

MOBILITY, CLOUDS, AND DIGITAL TOOLS USHER IN A CONNECTED WORLD

The Rise of the Global Village

Over the last decade, digital technologies have transformed the world. They've redefined the way people communicate, collaborate, shop, travel, read, research, watch movies, gather information, book vacations, manage their finances, and do so many more things. At the same time they've turned the modern enterprise upside down and remapped everything from sales to the how items flow through the supply chain. Today, the global Internet economy is somewhere in the neighborhood of $10 trillion US dollars.[1] By 2016, half the world's population—about 3 billion people—will use the Internet.

Mobility is at the heart of this revolution. Although mobile phones and laptop computers have been around for more than a quarter century—and personal digital assistants (PDAs) such as the Palm Pilot were available in the 1990s—it wasn't until Apple introduced the iPhone in 2007 that a single device could deliver a high level of functionality and a wide array of features in a near perfect form factor. The introduction of the iPad in 2010 provided further evidence that the mobile era had truly arrived. Suddenly it was possible to interact and transact in new and powerful ways. Today a typical smartphone that costs a few hundred dollars delivers more processing power than the $150,000 Apollo Guidance Computer that sent astronauts to the moon.

Approximately 6.8 billion mobile phones are now in use worldwide. In many developing nations, mobile phones represent the only way to venture online. Approximately 1.5 billion of these devices are smartphones and the number is rising rapidly. More important, the total number of mobile devices is now pushing beyond 2.5 billion. According to IT consulting firm Gartner, mobile devices now account for more than 50 percent of online activity. The post-PC era—a term attributed to MIT computer scientist David D. Clark in 1999—has clearly evolved from a futuristic concept into reality. According to The Boston Consulting Group, by 2016 mobile devices will account for four of five broadband connections.[2]

Smartphones—as well as tablet devices—are altering the way people access the Internet and share data. They're also creating new challenges and opportunities for businesses, educational institutions, government agencies, and others—as organizations attempt to harness social media and real-time data feeds. At the same time, mobile technology, along with cloud computing, is introducing new ways to manage connected devices. The web of connectivity and interconnectivity is an order of magnitude more powerful than anything that has come before it. The technology is a nothing short of revolutionary.

An iPhone or Android phone can now serve as the remote control for home theater equipment, operate thermostats, manage smart appliances, and interact with Internet enabled bathroom scales, baby monitors, automobiles, exercise and activity tools, heart rate monitors, and much more. Smartphones are capable of tracking fleets of vehicles and equipment, determining whether machines are operating correctly, and keeping track of children and pets. What's more, ports on iPhones, iPads, and other devices, connect to external devices and sensors, which further expands functions and capabilities. These days, the ideas are limited only by creativity and imagination—along with the existing technical limits of sensors, software, and batteries. But even these boundaries are rapidly disappearing as research advances and breakthroughs occur.

Into Thin Air

The idea for a mobile communication device extends back more than a century. In the late 1930s, the US military began to use radios, also known as "walkie-talkies," that weighed about 25 pounds and worked at a range of approximately 5 miles. In 1946, Chester Gould introduced the two-way watch radio in his Dick Tracy comic strips. The wristwatch communicator became a mainstay of the strip—and it clearly captured the public's imagination. Then, during the 1940s, researchers at Bell Labs, including Amos Joel Jr., W. Rae Young, and D. H. Ring, engineered a system that allowed callers to talk and exchange data while in motion. The technology allowed a communications device to connect to different cell towers—and switch towers based on location.

AT&T, then Bell Systems, introduced the world's first mobile phone service in St. Louis, Missouri, on June 17, 1946. The offering initially attracted only about 5,000 customers who placed about 30,000 calls per week.[3] The system wasn't exactly convenient, at least by today's standards. An operator had to connect the calls and it wasn't possible to use the service in other areas. The phones weighed 80 pounds and the service cost $15 per month, plus 30 to 40 cents per local call. Remarkably, only three subscribers could use the system at any given moment.

It wasn't until the 1960s that Bell Labs engineers Richard Frenkiel and Joel Engel assembled the computers and electronics to move beyond basic radio-based communication. Then, in 1973, Motorola engineer Martin Cooper placed the first call from a modern mobile phone on the streets of New York City. The device, which weighed in at slightly over 2.4 pounds and had a battery life of only 20 minutes, resembled a giant brick with an antenna sticking out. Yet another decade passed before mobile phones began to filter into the commercial marketplace. In 1979, NTT launched cellular service in Japan, Scandinavian countries commenced service in 1981, and the United States introduced service in 1983. The catch? The first widely available phone, the Motorola DynaTAC, had a price tag of nearly $4,000.

It wasn't until the 1990s—with the advent of modern cellular technology and smaller and lighter handsets—that mobile phones began to filter into the mainstream of society. The first attempt to build a digital smartphone came from IBM in 1993. The tech giant introduced Simon, a mobile phone, pager, fax machine, and PDA packed into a single device. Simon offered a number of features, including a calendar, address book, clock, calculator, notepad, and email. It introduced a touchscreen and used a stylus and QWERTY keyboard for input. Nokia, Erickson, and others, soon followed with the predecessors to today's icon-centric devices.

Then, in March 1997, the Palm Pilot entered the picture. Although Apple and others had previously introduced PDAs (Apple's Newton introduced the concept in 1992), the Palm device emerged as an overnight sensation. Among other things, it allowed users to store important personal data on the device and it synced with software on a computer. It also allowed users to add apps and additional features. Some models later included a modem and the ability to connect to the Internet. For the first time, users had a viable alternative to paper with pixels at their fingertips.

Alas, none of these solutions—or any of the smartphones available in the early 2000s—provided the type of point-to-point connectedness that we take for granted today. In fact, by current standards, they were clunky and at often frustrating to use. Internet connections were slow and sporadic, software didn't always deliver as advertised and the general interface was convoluted and confusing. These devices could best be described as the Model Ts of the mobility world. Although it was possible to connect some of the time, and in some ways these were hardly connected devices in the way we think about it today.

Nevertheless, the foundation had been laid for a connected mobile computing device. As cellular networks improved and Wi-Fi became more ubiquitous, pieces of the connectivity puzzle filled in. Smartphones soon sported chips that enabled both cellular and Wi-Fi connectivity. Then, with the introduction of the iPhone, adoption skyrocketed. It was possible to send and receive messages, view alerts and notifications, post on social media, and use apps to scan documents, exchange business cards, record audio, snap photos, read barcodes, and submit various types of data. New features and capabilities that were previously concepts suddenly became very real.

At the same time, cloud computing introduced better ways to sync and exchange documents, photos and data across different devices. Suddenly it was possible to use electronic boarding passes for flights, barcodes for hotel reservations, and digital wallets to pay for items ranging from a cup of coffee to an item at a garage sale. Meanwhile businesses began migrating from barcodes and manual inventory systems to RFID to tag and track pallets, vehicles, equipment, tools, and more. Some looked to the technology to boost efficiency within the confines of a factory or warehouse. Others turned to RFID to manage the supply chain more efficiently.

RFID is more than a tool for reducing costs and pushing up profits, however. It builds a bridge between the physical world and the virtual world. By attaching a small tag to an object (or installing a chip into a device)—either a tiny passive transponder using electromagnetic radiation or a battery powered passive or active tag that relies on UHF radio waves—and setting up an RFID reader, anything and everything can be connected to the Internet. Today RFID technology is used for toll collections, contactless payment systems, tracking animals, managing baggage at airports, embedding data in passports, following runners in a marathons, and tracking golf balls via a smartphone app.

How Mobile Technology Changes Everything

The ability to tag physical objects and transform anyone carrying a smartphone into a potential data point has remarkable and far-reaching implications. It's not evolutionary; it's revolutionary. The capability to extract data from a wide array of objects and devices in fact helps humans analyze things and gain far better insights. Instead of making educated guesses, it's possible to tap into data and analytics in order to understand patterns, trends, and behavior in a more thorough and comprehensive way.

One thing that distinguishes connected devices is that they can continuously report about usage, operating behavior, conditions, and other information. Simply put: they generate a lot of data that can be analyzed and acted upon. Combine human input with machine input and the stakes increase even more. The ability to pull data from social media, use crowdsourcing techniques, and include data from sensors, creates entirely new wrinkles and possibilities. With automation, rules, analytics, and artificial intelligence, there's the ability to achieve far greater intelligence about the world around us.

To be sure, mobile technology establishes the connection points—think of it as a central nervous system—between anything and everything on the planet. Smartphones and other handheld devices, RFID tags, sensors embedded in machines and even inside a body, and

The ability to tag physical objects and transform anyone carrying a smartphone into a potential data point has remarkable and far-reaching implications.

microchips built into objects offer a profoundly different way to measure and manage things that were previously indiscernible. Mobile technology also eliminates the time, expense, and complications associated with wiring or retrofitting buildings and houses. As broadband Internet and fast cellular networks blanket major swaths of the planet, the imitations on how data are collected, shared, and used are disappearing rapidly.

Yet mobile devices and networks alone could not produce the Internet of Things. Moving data from device to database and across vast computing networks that span countless individuals and businesses is a potentially complex, expensive, and cumbersome task. Just as a highway system requires more than roads and signs—an entire infrastructure of gasoline stations, cafés, motels, and other amenities must exist—the IoT requires systems, software, and tools for supporting everything. Without these components, it's merely a disparate collection of technologies that achieve limited functionality.

The intersection of various technologies with mobility—including cloud computing, social media, and big data—ratchets up the stakes. Ultimately each technology feeds into the others, and together, a far more powerful and expansive platform is born. It's something akin to a 1 + 1 = 3 equation. Indeed putting the IoT to work means not only understanding the ways devices can connect to one another but how networks and entire ecosystems of devices can alter data streams and create value. "The underlying technologies for the Internet of Things already exist. A large part of the puzzle is understanding how to fit together all the pieces in an appropriate manner," observes John Devlin, a practice director at ABI Research.

Nicholas Carr, author of the book, *The Big Switch: Rewiring the World, from Edison to Google*, points out that the introduction of widespread and inexpensive electric power in the early 1900s had implications that rippled into numerous corners of business, commerce and society. For instance, the landscape of cities began to change radically as elevators made it possible to build massive skyscrapers. The look of urban environments also changed as signs went up and shops could stay open after the sun went down. Likewise mobility and cloud computing create entirely new possibilities and introduce similar changes.

A Clearer View through Clouds

Today various types of computing clouds have become commonplace, and virtually no corner of the Internet or computing has gone untouched. Many describe the environment as utility computing because it's possible to switch on and off services at a moment's notice. What's more, it's possible to adjust usage patterns dynamically and in real time. But cloud computing also ushers in another reality:

the ability to process, route, and synchronize data far more efficiently than ever before. It would be virtually impossible for any one organization or government to build a data storage infrastructure capable of supporting the IoT. In addition, through the use of application programming interfaces (APIs)—essentially small programs that link applications—it's possible to build a far more flexible and automated environment. This software allows different devices and systems to speak with each other—even when they rely on different standards or protocols.

Although the term "cloud" is used broadly and in a different context for different situations, it essentially refers to a distributed computing environment that operates over an extended network, such as the Internet. Typically a collection of computers located on the Internet serves as a platform or service for users. This can take the form of software, hardware, and various services, including storage, that are delivered through the Internet or a private network. While the concept isn't new—the idea of hosted or managed services extends back to the 1950s through the concept of time sharing—radical advances in processing power, bandwidth, and software development have redefined the space over the last few years.

A good example of how the Internet of Things is taking shape—and how mobility and clouds play a role—is visible in a new wave of fitness devices. For years, runners, walkers, bicyclists and other fitness buffs, who wanted to track their progress had to either log their performance data with pencil and paper or purchase a device that could capture the steps and distance—and in some cases include the route using a built-in GPS. In recent years some devices could sync the information to an application or website using a cable or wireless technology such as Bluetooth. While these devices connected to the Internet, they were a crude version of what's possible within the Internet of Things.

During the last few years, a new breed of fitness devices has taken athletic performance—and tracking—to an entirely different level. For example, Fitbit wristbands track steps, calories, floors climbed, and active minutes via built-in electronics, including an accelerometer and an altimeter. Some models also track nighttime sleep patterns. These devices—which provide an organic light-emitting diode (OLED) readout—periodically connect to a smartphone or computer via Bluetooth and upload data to the cloud. There it is analyzed and the user receives insights in the form of charts, graphs, and other data visible at a website and through a mobile app.

Yet the power of the device extends beyond a dashboard. For one thing, the software integrates with other apps and sends data to them. This makes it possible to plug in data from Internet connected treadmills, exercise bikes, and other equipment. It's also possible to use heart rate monitors, other smartphone apps that track walking and running routes, and apps that track food and calorie intake. It's even possible to compete in fitness leagues with other users, track weight loss, and study wellness in ways that wouldn't have been imaginable only a few years ago.

What's remarkable about these capabilities isn't just the technology that measures and records activity in such a detailed and comprehensive fashion. It's the ecosystem of services and apps that connect to the Fitbit and similar devices. The result is an reasonably accurate motion picture of personal activity taking place during the entire day—from movement to eating habits and nutrition to sleep. A computer reads all the data from a collection of devices and apps, plugs the data into an algorithm, and delivers highly detailed results and analysis in real time. Without mobile technology, cloud computing and connected systems, all this wouldn't be possible. A user would be left with islands of data that provide limited insights.

Things Get Social

Mobility and clouds are fundamentally rewiring interactions and transactions in other ways. For instance, over the last decade, social media has matured from a novel idea into a mainstream phenomenon. At the beginning of 2014, more than 1.3 billion people used Facebook every month, and 680,000 of them relied on mobile devices to connect at least part of the time. Meanwhile Twitter boasts nearly 675 million users and about 58 million tweets per day.

These sites, and a bevy of others, are more than the sum of random posts. They provide a real-time glimpse into behavior, trends, and attitudes in areas as diverse as politics, entertainment, fashion, and consumption. The use of these data creates a connected world with entirely different touch points "In the past, a company would sell a product and it would disappear into a black hole. There was no way to know what anyone did with it or what other marketing opportunities existed," points out Glen All-mendinger, president of technology and business development at consulting firm Harbor Research. Today, in using social listening techniques, it's possible to view patterns that would have previously flown under the radar.

Increasingly these social media analytics applications use algorithms to tune into a growing universe of factors, including the number of hits or visits at a site or page, the number of unique visitors, the tone of comments, search engine rankings, click-through data, online discussion shares, the number of influential friends or followers a person has, changes in attitudes or sentiment within a person's social sphere, and myriad other factors. Additionally sites now combine human input with phone data. They use timestamps, check-in data, and geolocation data to better understand how consumers shop, dine, and travel. In every instance, these capabilities wouldn't be possible without a smartphone or tablet packed with sensors and real-time communication capabilities.

Following the Crowd

The human element also comes into focus in the crowd-sourcing arena. Areas such as health care are among the greatest beneficiaries of mobile technology and big data. In the physical world, understanding how an infection spreads and how people behave is highly variable and, at times, wildly unpredictable. Making sense of treatment methods adds layers of complexity to an already onerous task. But, by using smartphones, clouds, crowdsourcing, and big data analytics, it's possible to turn data collection upside down and inside out. Researchers now rely on these tools to examine everything from the spread of ordinary viruses to how eating and exercise impact obesity and health care costs. Emerging crowdsourcing models, such as CrowdMed, allow health care professionals to outsource an inquiry to other experts and receive answers within minutes or hours.

Crowdsourcing and the IoT have the potential to reach into a vast array of areas and touch people's lives in profound and different ways. "Technological advances ... are breaking down the cost barriers that once separated amateurs from professionals. Hobbyists, part-timers, and dabblers suddenly have a market for their efforts [as organizations] tap the latent talent of the crowd," points out Jeff Howe, who coined the term in 2006 and authored the book, *Crowdsourcing: Why the Power of the Crowd Is Driving the Future of Business.*[4] Howe sees crowdsourcing as a way to connect to knowledge and expertise that previously couldn't be tapped or analyzed.

Moreover city governments are introducing apps that allow citizens to report potholes and other problems via their smartphone. Relief agencies are using crowdsourcing to better understand how to focus aid and resources. For instance, Ushahidi, a software platform introduced in 2008 and collaboratively written by developers in Kenya, Ghana, South Africa, Malawi, the Netherlands and the United States—enables volunteers throughout the world to map everything from natural disasters to political turmoil. The result is sophisticated mashups with real-time visualizations, geospatial representations, and sophisticated crowd-mapping features.

All these capabilities redefine the traditional method of collecting data and putting it to use. Thanks to the Internet and increasingly low-cost technology—including smartphones—the barriers to connectedness and connectivity have dropped—along with the costs previously associated with gathering data from thousands or millions of people or devices. A process that would have in the past required paper, surface mail, and months of tabulation can now occur in moments—and results adjust dynamically as conditions or behaviors change.

Big Data = Big Results

Not surprisingly, various chips and sensors as well as human input from a smartphone or tablet generate vast amounts of data. Combined with existing sources of data—many organizations have legacy databases and records extending back decades—a vast new frontier of data exploration exists. Overall, data volumes are expanding at somewhere between 50 and 60 percent annually while mobile data traffic is growing at an annual rate of about 61 percent, according to networking giant Cisco Systems.[5] By 2020, International Data Corporation predicts that 40 zettabytes of data will exist worldwide (for perspective, 1,000 terabytes equals 1 petabyte; 1,000 petabytes equals 1 exabyte; and 1,000 exabytes equals 1 zettabyte. A single zettabyte would total about 250 billion DVDs, which is more than 35 years of nonstop viewing of high-definition video). That's six terabytes of data for every living person—roughly equivalent to three million books per person.[6]

Although big data has become a buzzword, it's a valid concept that centers on collecting, storing, and using datasets generated from both structured data (which resides in a database) and unstructured data (which exists outside a database), typically in the form of messaging streams, text documents, photos, video images, audio files, and social media. Doug Laney, now an analyst with Gartner, provided a short but effective explanation of big data back in 2001.

Not surprisingly, various chips and sensors as well as human input from a smartphone or tablet generate vast amounts of data. Combined with existing sources of data—many organizations have legacy databases and records extending back decades—a vast new frontier of data exploration exists.

He argued that it has three primary components: volume, velocity, and variety. Volume refers to the amount of data, velocity refers to the speed at which data are generated and put to use, and variety refers to the breadth of data that now exists.

Some disciplines—namely fields such as astronomy, meteorology, oil and gas exploration, and engineering—have long relied on huge datasets to solve problems and build models. The IoT exponentially increases the number of data sources along with the volume, velocity, and variety of data. Suddenly it's not only about computers collecting and generating data and storing the data in tidy databases. The IoT encompasses satellites, parking meters, vending machines, television sets, point-of-sale terminals, gas pumps, food packages, household appliances, light switches, restrooms, and supermarket shelves. It includes anything that can stream data to clouds and real-time analytics systems.

The challenge, moving forward, is to identify the right data and put datasets to use effectively. The ability to sift through big data and harness it will determine whether connected devices deliver on their full promise. To be sure, as Laney's three-Vs grow in importance—in large part due to digital convergence and the Internet of Things—business will find that it's critical to boost their velocity of analysis and velocity of action. They will be forced to move in swifter and smarter ways.

While this emerging wave of technology is introducing far more holistic and detailed ways to understand our world, the combination of sophisticated social listening systems, crowdsourcing models, and connected sensors and devices will deliver a far more granular level of analysis. It will be possible to improve the accuracy of weather forecasting models, develop a far more agile manufacturing model based on rapid innovation, use data to build better products, market these products more effectively, introduce new clothing lines or restaurant items in short order, and radically alter the way businesses interact with consumers.

Focus on the Future

One thing is undisputable: mobile devices will become even smarter in the months and years ahead. Today's smartphones can already "hear" and "feel" at a basic level. They possess built-in microphones, cameras, GPS chips, accelerometers, gyroscopes, and other sensors that can act and react to a wide range of environmental factors and conditions. Together, they create far more intelligence in the device and transform it from a phone into a multifunction computer that transforms our world.

In the not-too-distant future, smartphones will gain a sense of smell and taste—and become more context-aware.

Not only will this make phones smarter by eliminating rings, dings, and buzzes at a theater event or while taking a nap, it introduces even more advanced capabilities. For example, a phone that includes a temperature and humidity sensor and connects via Bluetooth technology to heart rate and blood pressure devices would provide deeper insights into athletic performance and overall health. It would also deliver more granular block-by-block data to weather forecasters, who could develop far more accurate prediction models based on highly detailed data resolution levels.

These concepts are now within the realm of possibility. San Francisco-based Adamant Technologies is currently developing a small processor that digitizes smell and taste. The system uses about 2,000 sensors to detect aromas and flavors. This compares to about 400 sensors in the human nose. The system would detect when a person has bad breath or is intoxicated and over the legal limit to drive. A digital nose in a smartphone could also one day detect underlying medical conditions or rancid food.

Moreover, if public health officials have access to these type of data—through crowdsourcing or automated data collection methods—it's suddenly possible to identify contaminated meat or other foods. If these packages are tagged using RFID, manufactures and stores could identify an affected lot and pull them from the shelves immediately, thus reducing the risk of widespread sickness. Likewise a phone that's able to sense could allow consumers to "feel" fabrics and textures while online. And apps could deliver augmented reality (AR) by allowing a person to hold up a phone's camera to an object—anything from a tree to a Mayan pyramid—and instantly view information about it.

Wearable technology—smart watches and bands, smart glasses such as Google Glass, and smart clothing—are already taking shape. These devices will extend and enhance the Internet of Things and make data more accessible. They could also reduce distraction and the need to pull a phone from a pocket or purse in order to check it. Electronic textiles and wearables also have the potential to sense body functions and even detect heat, high levels of ultraviolet light and chemicals, allergens and toxins in the environment. In fact Nike, Adidas, and others, have begun to embed sensors in shoes and clothing.

But the possibilities don't stop there. Using Bluetooth, near-field communication (NFC), RFID, and other wireless technologies, researchers are also exploring the use of nanosensors and optical fibers to peer inside collapsed buildings, industrial machinery, and the human body. There's further a growing focus on networks of smart objects or sensors—possibly numbering in the millions or billions—that interact with each other and behave in a context-aware manner. This might allow an army of package delivery drones to drop off orders in minutes and operate in the most efficient way possible. It might enable the development of smart tools or smart vehicles that ensure they're used the proper way and won't allow a user to exceed safety boundaries.

We'll examine these developments and other emerging technologies in the forthcoming chapters. Suffice it to say that a connected future is beginning to reveal itself—and mobile technology is the sun around which all the other technology planets orbit. The growing array of connected devices and systems—particularly in the consumer space—promise to change the way we live, work, and interact in the most profound way imaginable. We have only begun the journey.

THE INDUSTRIAL
INTERNET EMERGES

A New Model Takes Shape

At the heart of the Internet of Things is the Industrial Internet. It provides the underlying infrastructure that supports connected machines and data. The term, which is generally attributed to manufacturing giant GE, refers to the integration of machines with sensors, software, and communications systems that enable the Internet of Things. The Industrial Internet pulls together technology and processes from fields such as big data, machine learning, and M2M connectivity.

Some refer to this connected business world as Industry 4.0, alluding to the fourth wave of disruptive industrial innovation (previous waves include mechanization, mass production, and the introduction of computers and electronics), or simply smart industry or smart manufacturing.

Some refer to this connected business world as Industry 4.0, alluding to the fourth wave of disruptive industrial innovation (previous waves include mechanization, mass production, and the introduction of computers and electronics), or simply smart industry or smart manufacturing.

Not surprisingly, different companies have introduced their own catchy monikers. For instance, IBM describes refers to a Smart Planet and Cisco Systems simply uses Internet of Things.

Regardless of the exact term that's used, the structural framework for this next step in technology and business is essentially the same. The Industrial Internet and Internet of Things share the same technology underpinnings and the same virtual space, though the former is considered a distinct entity or component of the IoT. But both share the common goal of blending and blurring the physical and virtual worlds—as well as the distinctions between human and machine—in order to generate far greater intelligence than any single machine or device can produce.

So far the Industrial Internet has revolved heavily around smart utility meters, vehicle and asset tracking, and optimizing the performance of plants, facilities, and machines. However, over the next few years, existing digital devices will mesh with machines in much deeper and broader ways. Additionally the Industrial Internet will serve as the foundation for a growing array of consumer devices and systems, which we will examine in the next chapter.

As McKinsey's *the Internet of Things* report notes[1]:

Business models based on today's largely static information architectures face challenges as new ways of creating value arise. When a customer's buying preferences are sensed in real time at a specific location, dynamic pricing may increase the odds of a purchase. Knowing how often or intensively a product is used can create additional options—usage fees rather than outright sale, for example. Manufacturing processes studded with a multitude of sensors can be controlled more precisely, raising efficiency. And when operating environments are monitored continuously for hazards or when objects can take corrective action to avoid damage, risks and costs diminish. Companies that take advantage of these capabilities stand to gain against competitors that don't.

Data Matters

At the most basic level the IoT and Industrial Internet are about data and extracting value from it. Today, thanks to pervasive computing and nearly ubiquitous networking, data bits and bytes travel to nearly every corner of the planet in real time. A growing array of devices—including desktop computers, laptops, tablets, and smartphones—serve as conduits for collecting, sharing and accessing rapidly growing volumes of data. Of course, connected devices—everything from insulin pumps in hospitals to lighting systems at home—ultimately depend on data to function—or provide feedback for making decisions.

Data scientists have coined a term, *value of perfect information*, which revolves around the ability to align data points, collection, and analysis in a way that delivers deep insights. Achieving this goal is incredibly challenging because it's extraordinarily difficult to gather all the data required for perfect information and then build an algorithm that takes into account all the variables in the right way. For instance, the ability to forecast weather is dependent on collecting data at a highly detailed and granular level, plugging in the relevant data, and making sense of the data through sophisticated algorithms. Theoretically, if scientists could put the right systems and software in place—and have access to enough computing power—they could produce 100 percent accurate forecasts.

At least for now, too many variables and limitations exist to achieve a perfect view of any complex event—whether it revolves around weather, agriculture, manufacturing, health care, transportation, or the stock market. So, instead of looking to build perfect models, data scientists are focused on building the best possible models using big data and analytics. This encompasses predictive analytics, which aims to identify or understand an event before it takes place. For example, this might allow a bank to know when a customer is likely to change institutions or identify a consumer who is considering a new automobile but hasn't yet begun to shop for it. It could also help organizations understand when a machine part is likely to fail or what products a person will likely buy in a store.

The data stream from connected machines and objects is growing exponentially. According to a report from data management company Wipro, *Big Data: Catalyzing Performance in Manufacturing*, a 6-hour flight on a Boeing 737 from New York to Los Angeles generates a whopping 120 terabytes of data that is collected and stored on the plane.[2] More important, all these data can be analyzed to reveal every aspect of the engine's performance and health.

Not surprisingly, data are becoming a valuable economic asset. Information technology consulting firm Gartner in fact predicts that information assets and data will appear on the balance sheets of corporations within the next few years. The emergence of data as a currency could impact stock valuations, merger and acquisition activity, and much more. This economic value extends beyond assets, however. McKinsey Global Institute estimates that big data could decrease product development costs in manufacturing by 50 percent or more.[3] The ability of analytics software to pore over huge numbers of data points could also spot quality or service gaps, trim operating costs, and fundamentally change the way organizations view investments in machinery and people.

Indeed the landscape is changing as organizations learn how to tap into big data and put data to use. Although databases, software applications and unstructured data streams already deliver a wealth of insights, these sources pale in comparison to the vast untapped data sphere that exists within the physical confines of our planet. Historically there's been no way to measure, collect, or process these data. The information existed beyond the limits of our vision, senses, and instruments in much the same way that radio waves, ultraviolent light, and other signals seemingly don't exist. They became relevant to humans only after we build devices and systems capable of detecting these electromagnetic waves.

The Internet of Things promises to ratchet up the number of data points by an order of magnitude. The combination of ubiquitous connectivity, low-cost sensors, and easy to deploy microelectronics now make it possible to connect just about anything and everything to the Internet. Suddenly milk cartons, roads, bridges, vehicles, trees, machines, medical devices, and power systems become data points. Of course, all the intersecting data create entirely new insights and opportunities.

Sensing Gains

Sensors are at the heart of the Industrial Internet. Over the last few years, radical advances in technology along with miniaturization have created new opportunities to sense things in the natural environment.

Today the list of data input points and connected systems includes things as diverse as geolocation and GPS devices, bar code scanners, thermometers, barometers, humidity gauges, vibration sensors, pressure sensors, gyroscopes, magnetometers, cameras, audio and video monitors, accelerometers, motion sensors, radar, sonar, and lidar. The latter is used by Google to operate its autonomous (driverless) vehicle fleet, now known as *Google Chauffeur*, which has traveled more than 700,000 autonomous miles without a technology-caused collision.

Yet, while sensors collect data, it also takes computers, storage systems, and software to manage and make sense of data. Connected systems—often relying on an application programming interface (API) to make data available by and for applications when and where data are needed (these small software components connect different devices and software programs, essentially defining interactions and how data exchanges take place)—enable back-end processing for things such as data mining, facial recognition, and translation systems. For instance, a system might recognize a person or use her facial expressions to suggest items when she steps into a store or allow a person to snap a photo of a sign or message in another language and receive an instant translation. It also introduces augmented reality, which lets a person snap a picture of a thing, say the Eiffel Tower, and receive information about it instantly. The translucent print appears as an overlay on the original image or on the display for smart glasses, such as Google Glass.

The possibilities are nearly endless and the potential benefits for a business are significant. The industrial Internet of Things, according to McKinsey consultants Michael Chui, Markus Löffler, and Roger Roberts, represents an entirely new wave of opportunity. As they wrote in a 2010 report called *The Internet of Things*[4]:

> The predictable pathways of information are changing: the physical world itself is becoming a type of information system. ... These networks churn out huge volumes of data that flow to computers for analysis. When objects can both sense the environment and communicate, they become tools for understanding complexity and responding to it swiftly. What's revolutionary in all this is that these physical information systems are now beginning to be deployed, and some of them even work largely without human intervention.

How does all of this play out on the front lines of business? Machine-generated data currently accounts for about 15 percent of the overall data organizations hold. However, the figure will likely rise to around 50 percent within the next decade. Intelligent assets—essentially devices equipped with sensors and connected to one another—will deliver parameter readings, usage information, operator behavior, and condition and health monitoring.

Within the industrial and commercial realm, the IoT will likely spawn huge gains. Even a 1 percent reduction in fuel costs or a similar improvement in capital expenditures of system inefficiency could produce savings in the tens of billions or hundreds of billions of dollars. The Industrial Internet could also spawn economic activity measuring in the tens of trillions of dollars.

A Connected World Changes Everything

Several key capabilities grow out of the Industrial Internet—and these components often overlap as organizations deploy the technology. Among them:

Location Awareness

It's now possible to track movements and motion through a growing array of cameras, sensors and satellites. This data increasingly defines our world. Digital cameras record geolocation data with photos, cellular towers time stamp the exact moment a user passes by with a mobile phone, card readers and transponder systems, such as E-ZPass, record when drivers travel past a tollbooth, and social media apps such as Facebook, Twitter, and Yelp rely on presence technology to record when and where a person posts a status line update or a check-in. At the same time, GPS chips and satellites identify exactly where planes, trains, and vehicles are located at any given moment.

While the Global Positioning System has been in place for more than two decades—the idea for a series of satellites orbiting the earth was envisioned back in the 1950s—it represented only a piece of the overall location awareness puzzle. Other key components include: computing devices with Media Access Control (MAC) addresses, which provide a unique machine identifier; IP addresses that track a device's location on a network such as the Internet; an Ethernet address that indicates a device's location on a local area network (LAN); and RFID tags and similar sensors that bridge the physical and virtual worlds.

Within this new IoT order, smartphones represent the last mile in geolocation. They enable data collection on an ongoing and constant basis through the use of GPS chips, cell tower triangulation, and, when these signals are too weak or unavailable because of a building or obstruction, local Wi-Fi data or Assisted GPS (A-GPS) technology. These rely on a variety of network resources to identify a user along with the person's location in order to communicate with the device.

Already real-time location systems (RTLS) are used in a number of industries and businesses. Among these:

• Navigation systems based on GPS as well as cellular technology are widely used for tracking trucks, ships, and airplanes as they move from one place to another.

• Fleet tracking systems allow logistics and transport firms to optimize routing, analyze driver efficiency, track speed and vehicle location and use better understand fuel and maintenance costs.

• Inventory and asset tracking technology, often incorporating RFID, identify physical assets or follow them through a supply chain. For more than a decade, retailers have used these systems at the pallet or case level to identify the location of goods in transit. However, retailers and others are now taking RFID to an item level. This makes it possible to build far more robust systems and introduce entirely new features and capabilities.

• Personnel tracking and authentication. RFID-enabled badges, smartphone apps using GPS and location aware services, and other tools, make it possible to know where a person is at any instant. The technology is widely used in secure facilities and labs, including government offices and military bases with strict authorization or access controls.

A good example of RTLS and its role in the Internet of Things is visible at Oregon Health Sciences University (OHSU). The Portland, Oregon, health and research institution tags assets ranging from infusion pumps to crutches so that they are easy to locate. In addition they are able to track performance data related to the device. This approach not only saves time that would otherwise be spent hunting down equipment, it helps ensure that devices are in working order. OHSU is now looking into tagging patients and clinicians to better understand where they spend time, how they move around within the facility, and how long patients wait in a room before a clinician arrives. "The technology provides insights into how we can operate more efficiently," says Dennis Minsent, director of clinical technology.

Another organization, Hawaiian Legacy Hardwoods, has morphed low-tech lumber and ecotourism with high-tech systems. Since 2010, the Honolulu-based firm has tagged more than 225,000 trees using passive RFID with GPS coordinates. A database contains information about the seed stock, feeding schedules, watering schedules, and other information. Over that time the company has added to its database and altered processes based on information obtained from the network of trees. "We can keep track of pretty much any event involving a tree. We simply scan it, record it and log it," says CIO William Gilliam.

Technology such as Apple's iBeacon introduces even more robust capabilities that could alter the way people shop. Once a retailer is able to track a person's route through a store and identify when and where he or she is hovering, it's possible to plug in past buying and behavioral patterns, run the data through analytics software, and determine whether it's productive to offer a coupon or incentive, and, if so, how much of a discount or how many bonus reward points to offer. The aggregate data collected from thousands or tens of thousands of shoppers can also help the retailer design a store better or optimize the layout of shelves and products to improve sales. The analytics software spots trends and relationships invisible to the human eye.

Enhanced Situational Awareness
Another way to use sensors is to embed them in the physical environment, including roadways, buildings, soil, plants, and the ocean. When hundreds or thousands of sensors connect to one another, it's possible to view data at a much higher resolution and understand relationships and patterns in a much more detailed way. Within a city a smart transportation network, for example, could route traffic with maximum efficiency and optimize traffic lights for maximum flow. This would not only speed commutes but also allow more vehicles to share the road at any moment.

But the technology is also yielding benefits in areas a diverse as agriculture and weather forecasting. Today farmers use sensors embedded in machinery and fields to dispense fertilizers and pesticides at more precise—and environmentally friendly—levels. Sensors monitor the moisture level of fields and switch on irrigation systems based on soil moisture and weather forecasts. Even cows, pigs, and other animals are getting connected. By some estimates there are already more than 14 million connected farms in the United States and Europe, and by 2020 there could be upward of 70 million connected devices. Companies such as OnFarm and Trimble are leading the data revolution by introducing systems that gauge everything from soil moisture and soil tension to pH levels and optimal fertilization patterns.

At the same time increasingly powerful computers, sensors, big data, and more sophisticated computer modeling and simulation tools are taking weather forecasting to a new and far more accurate level. Today's six-day forecasts are about on par with five-day forecasts a decade ago. In addition to satellites and conventional weather stations, researchers are deploying an ever-growing number of sensors in the physical environment. Ben Kyger, director of central operations for the National Centers for Environmental Prediction at the National Oceanic and Atmospheric Administration (NOAA), says that the goal is to increase the grid resolution to develop more accurate forecasts that extend further out in time. In fact, the National Weather Service has already experimented with social media input and crowdsourcing techniques to plug in weather data on a micro-block level.

Additional sensors and crowdsourcing techniques create new opportunities to obtain data at a higher resolution. Lloyd Treinish, an IBM Distinguished Engineer at the IBM Thomas J. Watson Research Center, which oversees IBMs Deep Thunder initiative (it strives to improve weather forecasting models through more sophisticated data collection and analytics), says that measurements from farm equipment, wind and moisture sensors in the soil and air, and data from smartphones and other devices, provide insights into everything from temperature and wind conditions to barometric pressure and humidity on a block-by-block level. With the right data and the right data points it's suddenly possible to ratchet up prediction capabilities and build far more useful and economically valuable models.

The data derived from a connected physical world could also extend to physical infrastructure and public safety. With connected bridges, tunnels, and roadways it's suddenly possible to understand when a structure is approaching a state of failure. This makes it much easier to prioritize risks—and repairs. What's more, with the right software and dashboard in place, it's possible to view data across an entire physical infrastructure. In other words, an agency or organization could clearly determine—based on structural data rather than opinions and politics—the real world risks and costs of fixing or ignoring a problem.

Today businesses and government are already using enhanced situational awareness to make to manage traffic. Law enforcement agencies in Los Angeles, New York, Memphis, and Santa Cruz are using data input and real-time analytics to pinpoint probable risk and assign officers and resources based on the emerging concept of predictive policing. Companies are installing a growing array of sensors—ranging from video monitors to listening devices—to detect problems and use data to better manage road construction, water management facilities, and aircraft manufacturing plants, to name a few.

Sensor-Based Decision Analytics
The Internet of Things also supports longer range, more complex human planning and decision-making. With enough computational power, the right sensors, and sufficient storage, it's possible to take data collection and analysis to a level that would have once seemed entirely unimaginable. For instance, McKinsey and Company points out that extensive sensor networks placed in the earth's crust could produce an entirely new level of observation and insight for drilling companies. A good example of how sensors and monitoring are changing things is apparent in how Swiss-based multinational oilfield service firm Weatherford uses RFID today. The company is able to gauge the condition of drilling equipment and determine when repairs and upgrades need to take place.

At The Swedish Transport Administration, known as Trafikverket, visual inspections of heavily used wagons and carriages has been replaced with electronic detection systems along 13,000 kilometers of track. Using RFID tags and readers as well as cellular and local area networks (LAN) to transit data to a central monitoring facility, technicians obtain data from upward of 150 wayside monitoring spots. They are able to detect overheated axel bearings, wheel damage, vibration problems, and other issues as trains whiz past at full speed. "We are able to take the equipment out of use before major damages occur," says Lennart Andersson, RFID project manager.

Aggregate data—used by retailers to better understand buying habits, manufacturers to understand equipment, or health care firms to more precisely predict behavior—will revolutionize every industry over the next quarter century. Cameras, video, audio, motion data, and other input sources will create new and improved algorithms, simulations, and modeling methods. Once environments and people are equipped with sensors, businesses, government, and others, are able to transform data from a snapshot that takes place every day to a motion picture that allows moment-by-moment adjustments, adaptation, and alterations.

The net effect could be profound. Sensor-based decision analytics provide immediate feedback about an event or situation but also provide a deeper view into usage and consumption patterns in real time. This in turn creates an opportunity for pay-as-you-go pricing as well as fee models that adjust dynamically to waning and waxing demand or other factors. Airlines have already begun to use a dynamic model to adjust pricing in real time—though today's models pale in comparison to what's possible in a highly connected business world.

In the near future a spate of monitoring devices could also measure fitness, health, and food consumption on a real-time basis and allow health insurance providers, for example, to base premiums and coverage levels on specific measurements from devices as well as conventional medical exams and lab tests. Using this model, those who volunteer to provide data and demonstrate a healthier lifestyle might receive a financial incentive, perhaps a lower monthly premium.

Automation and Controls

The final structural component to the Industrial Internet is the building of systems that use machine intelligence—some describe this as artificial intelligence—to automate processes and decisions. The ability to take humans out of the loop produces speed and efficiency gains that can radically redefine business, education and government.

During the last few decades, robots have taken over assembly lines and manufacturing. They rivet, spray, and weld their way through a variety of challenging —if not rote and dangerous—tasks. They also have moved into medicine—serving as surgical tools and providing limb replacements that increasingly mimic human motion. But they are now gaining sense, including vision and touch, along with a level of artificial intelligence, or AI, that allows them to operate autonomously. As the twenty-first century unfolds, they will become commonplace.

Advanced robotics and machine intelligence are likely to remove the human element from manufacturing and hard labor. They will likely redefine everything from package delivery and window washing to road repair and warfare. What's more, radical advances in machine intelligence will likely lead to robots and other systems that constantly analyze performance and learn to correct their own mistakes—as well as other machines and people. As networks of sensors feed waves of data to computers for analysis and algorithms and software become better at understanding and acting on the data in a contextual way, new and sometimes remarkable levels of automation and intelligence result. An industrial system or a robot can automatically adjust the way tools and machinery are used, the way chemicals and ingredients are mixed, or the way a company manages or maintains motors in jet engines or robots used for manufacturing goods.

In addition a vast network of sensors can provide immediate feedback about changing conditions. This is particularly valuable for managing limited or scarce resources, such as energy or water. A growing number of utilities are in fact introducing smart meters that track and display real-time consumption patterns. These meters also provide tools for viewing variable rates and taking advantage of nonpeak pricing. In the future smart utility grids would allow homeowners and businesses—including large data centers that consume vast amounts of electricity—to tap into sophisticated algorithms to optimize usage, save energy, and trim costs. Today smart thermostats connected

Advanced robotics and machine intelligence are likely to remove the human element from manufacturing and hard labor. They will likely redefine everything from package delivery and window washing to road repair and warfare.

to air-conditioning units already determine when to turn the fan on and off and how to mix outside and inside air optimally based on the temperature inside and outside the structure.

The same technology introduces rapid, real-time sensing of unpredictable conditions along with instantaneous responses guided by automated systems. This concept revolves around artificial intelligence, or AI. The goal? Mimic human thinking and decision-making but take the concept far beyond human capabilities. Today's automobile and airplane collision avoidance systems—which produce audible alarms and in some cases take corrective action—are an example of AI in action. However, in the future these capabilities will continue to advance. For example, armies of nanobot drones or robotic insects could be sent out to clean up an oil or toxic waste spill or explore a building collapse to find survivors after an earthquake.

A Connected Military

The Industrial Internet also serves as the foundation for an entirely different military and radically different warfare. In recent years, military drones have altered the face of combat and the way governments stalk terrorists. The United States, for example, has deployed unmanned air vehicles (UAV) in Afghanistan, Pakistan, Somalia, and other locales. Part of the appeal is cost: the American Security Project reports that military drones have a price tag of about $6.5 million compared to nearly $100 million for an equivalent fighter jet—without pilot risk. In addition drones are far less expensive to operate than manned aircraft.[5]

An operator, sitting at a remote location—in some cases thousands of miles away and connected through the Internet—operates the plane and pulls the trigger on gunfire, missiles, and bombs in a scenario that resembles a video game. But stand-alone drones are only the first phase of a more connected military. Over the next few years, vehicles, heavy equipment, medical devices, goggles, and much more will get connected. Systems will increasingly incorporate augmented reality—the overlay of data and computer-generated information with actual events.

Further out, far more advanced capabilities are likely to emerge. The US Defense Advanced Research Projects Agency, or DARPA, is exploring the use of robot armies—including insects—that would crawl, slither, slink, and fly behind enemy lines to accomplish the desired mission. This might include autonomous killing—something the United Nations and other groups are exploring both in a humanitarian and legal sense. But these devices could also tackle a variety of dangerous tasks, such as diffusing bombs or rescuing injured soldiers in the field. Moreover connected devices have enormous implications beyond the battlefield. They could assist with the collection of data and the ability to analyze it more effectively. This has repercussions in everything from intelligence gathering to predicting behavior and managing resources.

Making Connections Count

As an ever-growing number of industrial machines and components become part of the IoT—including a vast segment of legacy systems such as boilers; heating, ventilation, and air conditioning (HVAC); train and boat engines; and electrical systems—the nature, design, and workings of buildings, transportation systems, and factories change significantly. Embedded sensors and continuous connectivity make it possible to monitor food packages for spoilage, tires for wear and tear, and roofs for leaks. What's more, as robotics and nanotechnology intersect with the IoT (we will take a closer look at this topic in chapter 7), entirely new and often mind-bending capabilities are emerging, including networked autonomous devices that could handle dangerous construction and demolition projects.

Make no mistake, the Industrial Internet and larger Internet of Things provide deep insights into a vast array of physical systems. The impact of the technology is already significant. Within enterprise supply chains, sensors provide immediate feedback about the condition and location of goods. End-to-end monitoring creates an entirely different type of digital business that's more agile and cost efficient. It's possible to innovate and bring products to market faster, source materials and components more effectively, route products to market more efficiently, and provide a much higher level of customer service.

The combination of technology and systems also opens up entirely new business frontiers. In December 2013, Jeff Bezos, CEO of online retailing giant Amazon.com, announced plans to operate a fleet of drones and deliver packages on demand within a few years. We'll examine the social implications in chapter 6 but one fact is perfectly clear: drones will force entire industries to undergo radical change and, if the idea catches on, drone deliveries could fundamentally change the way consumers buy and use goods—and even recycle old products. 3D printing, which manufactures products on demand, will likely revolutionize things further. It already is used within numerous industries

Emerging pricing and usage models rely on the IoT and a connected physical environment that manages data in real time. Pay-as-you-go insurance is only the start. In the aviation industry, jet engine manufacturers increasingly retain ownership of their products and charge airlines based on the actual engine use measured by thrust. In many cities it's now possible to rent a bicycle or car by the hour through pay-as-you-go services. Automobile rental service, Zipcar, for example, directs a person with a smartphone to the nearest vehicle. An RFID transponder unlocks the car and a black box inside the car transmits data back to a server via a wireless link (though the company does not track the location of a customer for privacy reasons). The vehicle is also equipped with a kill switch in the event it's lost or stolen.

In the public health arena, connected devices combined with crowdsourcing techniques are changing the way experts view and treat outbreaks, such as the flu. The ability to view visualizations in real time and watch patterns change provides insights into how a virus is spreading and where there's a need for additional medicine and resources. Additionally increasingly sophisticated computer simulations can run different outbreak scenarios and show how different approaches impact threats as diverse as HIV and poison gas.

At MIT, researchers in the Department of Civil and Environmental Engineering, led by associate professor Ruben Juanes, are applying smartphones and crowdsourced data to better understand the role of the forty largest US airports in influencing the spread of contagious diseases. The project could help determine appropriate measures for containing infection in specific geographic areas and aid public health officials in making decisions about the distribution of vaccinations or treatments in the earliest days of contagion.

In order to predict how fast a contagion might spread, Juanes and his team are examining variations in travel patterns among individuals, the geographic locations of airports, the disparity in interactions among airports, and waiting times at individual airports. Juanes, a geoscientist, has tapped past research on the flow of fluids through fracture networks in subsurface rock to build an algorithm for the current endeavor. What's more, the team plugs in cell phone usage data to understand real-world human mobility patterns. The end result is "a model that's very different from a typical diffusion model," he says. Without the Internet of Things, all of this wouldn't be possible.

It's safe to say that the industrial Internet represents a giant leap forward. M2M connectivity is the foundation for the next generation of government and business. The ability of machines to talk to each other over a network— a process called telemetry—takes things to an entirely different level. It will lead to faster and better decisions and far more automation. It will support a growing array of consumer devices and services. Yet, in order to tap into the full potential of the IoT, organizations must learn how to integrate systems, devices, and data and put them to use amid a backdrop of rapidly growing security risks and privacy concerns. There are significant issues surrounding safety and automation as well as technical challenges that could derail—or at the very least slow down—adoption and results.

CONSUMER DEVICES GET SMART

A World without Wires

The typical house now has somewhere around 75 electrical outlets. Take a look at these receptacles—and count both the devices that are hard-wired to electrical systems and the machines we plug in periodically—and you might find somewhere in the neighborhood of 200 to 300 objects and devices scattered about. These range from ovens and central heating units to vacuum cleaners, lamps, iPads, and battery rechargers. By now, it's obvious that we've become highly dependent on a mind-bending array of machines to power our daily lives.

At some point in the past, every device represented a leading edge breakthrough. Each beckoned with the promise of a better or more convenient future. Washing machines removed the manual drudgery of pounding clothes on rocks or rubbing them with sand and soap. Lamps

Connected devices change the way we think about products and things, and they drive enormous changes in behavior as well.

provided a way to incorporate lighting in a more targeted and task-centric way. Toasters made it easy to brown bread without starting an oven or lighting a fire. Radios made it possible to hear news and information in minutes rather than the next day in a newspaper. And electric hedge clippers allowed gardeners to trim bushes and shrubs in seconds or minutes rather than hours.

Today we take these devices—and many others—for granted. They're simply part of our daily life. Of course, over the years many of these machines have become far more advanced, and infinitely better. Washing machines and other appliances use circuit boards to clean better, automate tasks, and save energy. Lamps provide dimmers to adjust the brightness. Toasters provide automatic settings and even sensors to detect when a bagel is just how we want it. Meanwhile radios are built into computers and hedge clippers offer far better designs with advanced safety features.

The advance of consumer technology has been nothing short of breathtaking—even if we don't always recognize it. What's more, the introduction of consumer electronics—and in recent years powerful computational capabilities built into devices —has profoundly changed the way we watch movies and television, communicate, shop for goods, gather information, and navigate a dizzying array of other tasks. It's safe to say that the world is a much better place as a result of technology innovation. It has delivered leisure time and helped drive societal gains. It has made our machines and cars safer, our medicine better, and created comforts that past generations could only fantasize about.

Some sociologists and cultural anthologists, including Alvin Toffler and Daniel Bell, have advanced the idea that we're heading into a postindustrial age that places an emphasis on information and services rather than the mere consumption and use of goods. There's plenty of evidence to support this notion. According to various market research reports, consumers now have about seven connected devices per household in the United States. However, the Organization for Economic Cooperation and Development (OECD) estimates that the figure will reach twenty by 2020.[1] What's more, the technology surrounding these devices is increasingly important. Market research firm NPD Group found that 88 percent of mobile device owners are now aware of home automation systems.[2] What's more, a growing percentage of individuals with smartphones, e-readers, Blu-ray players, and other devices, say that Internet connectivity and the ability to view content—in some cases across multiple devices—is a primary appeal.

Connected devices change the way we think about products and things, and they drive enormous changes in behavior as well. A quarter century ago, the primary way to view a movie was to head to the local theater or Cineplex and plunk down several dollars for each ticket. Today we purchase or rent movies from streaming media players attached to television sets or watch them wirelessly through tablets, smartphones, and gaming consoles. We view films and listen to music downloads on airplanes and in coffee shops. No less significant: social media reviews and recommendations increasingly influence thinking and buying decisions.

Connected devices translate into connected people—along with entirely different relationships among groups of people. Yet these human connections, however important and profound, are only a piece of the overall IoT puzzle. An individual device or thing connected to the Internet increases the power of that particular device—and often adds substantial value for the person using it. However, the ability to connect devices into a vast network—essentially the Internet of Things—increases the possibilities and capabilities exponentially.

For instance, a light switch that's Internet enabled not only allows a homeowner to program on and off times with a smartphone and manually control it from the same phone, it can be connected to software that analyzes electrical consumption across all lights in the house and, by offering recommendations, save money. Scaling up even further, the same data could be used by a utility to better understand consumption patterns and establish rates and incentives that drive more efficient usage patterns across

Connected devices translate into connected people—along with entirely different relationships among groups of people. Yet these human connections, however important and profound, are only a piece of the overall IoT puzzle. An individual device or thing connected to the

Internet increases the power of that particular device—and often adds substantial value for the person using it. However, the ability to connect devices into a vast network— essentially the Internet of Things—increases the possibilities and capabilities exponentially.

a customer base. It's not difficult to identify similar possibilities in many other industries, including automobiles, health care, and financial services.

Moreover attaching RFID tags and other sensors to various objects and packages introduces remarkable capabilities. Suddenly it's possible for a kitchen pantry to recognize when the supply of rice or salsa is low. A refrigerator can determine that the bread or butter has run out and it's time to buy more. A bathroom cabinet can alert a homeowner to buy more toilet paper or toothpaste—and even automatically add the items to a shopping list. Then, when the consumer steps into a grocery store and approaches the aisle with the desired product, he or she receives a smartphone alert or notification—and perhaps even a coupon.

Of course, a greater number of connected devices translate into more data intersection points—and far more impressive possibilities. Realistically we've only begun to enter the age of connected devices. Although home networks and Wi-Fi have been widely used for more than a decade—and fast cellular connectivity is increasingly common—the platform and infrastructure for supporting all these devices is only now beginning to mature. Too often in the past, various systems and devices did not communicate or play nicely with one another. What's more, without clouds that make sharing and syncing data far less complicated, fast and seamless data sharing simply wasn't possible.

Today the pace of innovation is accelerating rapidly and digital technologies are maturing. As data platforms take hold, analytics advances, clouds become a standard part of information technology, mobile applications grow in power and sophistication, and prices for RFID and other sensors plummet, the foundation for the Internet of Things is taking shape. Clearly, our world will never be the same. We are entering a new era that promises to revolutionize everything we do.

Connections Count

The idea of connected devices is nothing new. For decades, it's been possible to plug headphones into an audio jack on a stereo system or portable CD player and extend listening to a more private sphere. It's also been relatively simple to place a timer between a lamp and an electrical outlet and control when the lights turn on and off, and use remote controls to manage electronic devices. In the PC era, USB ports made it relatively simple to connect an array of peripherals—including external hard drives, digital cameras, digital audio recorders, headsets and microphones, blood pressure monitors, musical instruments, and scores of other devices.

There's no question that adding peripherals and components to a device or tool can add to its value. But a plugged in device is just that—a single object connected to another single object. This model delivers limited features and capabilities. Moreover there's no ability to use a device—such as a light timer—in a more sophisticated way. Making matters worse, the interface for many of these devices is clunky at best and perplexing at worst. Many even require manual programming and constant reprogramming. By contrast, an Internet connected light switch can check the Internet daily for sunrise and sunset information and automatically adjust to changing light conditions. It also allows a user to set and reset rules easily from a smartphone, and use text messages through a service such as IFTTT to toggle on and off.

Indeed connected devices and systems have become far more sophisticated over the last few years. Thanks to better user interfaces, improved software, easy remote access, improved technology standards, and consumers who are more comfortable using devices, a platform for connectedness and interactivity has emerged. These technology improvements—along with faster semiconductors, GPS, accelerometers, and other sensors—have eliminated physical layers of hardware and underlying code. That in turn has driven down prices and further fueled demand for connected devices.

It takes only a glance at remote controls to see this trend is evolving. Once, consumers accumulated a remote control with the purchase of every electronic device. That led to universal remotes that consolidated functionality for multiple machines in a single device. But manually programming the codes for a TV, DVD player, radio tuner, media streaming unit, and more, into a single remote can prove arduous—and sometimes nothing short of torturous. However, newer software-based systems reside on a smartphone or tablet. Setting up the device is fairly simple and straightforward. A user enters the manufacturer and model number of the unit, the software checks a database and automatically programs the remote—and then sets it up all the devices to operate in harmony with one another.

Smarter software and systems—and more sophisticated and targeted algorithms—have not only translated into tools that dramatically reduce the number of steps required to set up or use a device. They've also introduced entirely new features. For instance, in the early 2000s, streaming media players began to appear and digital video recorders such as TIVO surged in popularity. These devices demonstrate just how powerful a connected device is and how much it can alter the course of entire industries. Within a few years, consumers were time shifting, viewing and watching shows on demand. Soon Internet-based streaming services and radio stations popped up to address the burgeoning interest in content delivered on an *anytime, anywhere* basis. As the audience for traditional radio and television diminished, entire business models began to undergo fundamental change.

Today, thanks in part to the emerging Internet of Things, consumer behavior and consumption patterns are changing markedly. Consumers have already tossed

CDs for digital music downloads and streaming music services. They're closing the book on hardcover and paperback books as they adopt Kindles, Nooks, and iPads with Apple's iBook app. Video stores have vanished. In addition magazines are going electronic and paper maps are folding into the history books. Along the way, TV viewership has plunged from peak levels—today's shows typically claim only about 40 to 50 percent of the viewers their counterparts did a couple of decades ago—while newspaper ad revenues have declined by about 50 percent since 2005.

In other words, a connected world is driving a level of disruption that is unparalleled in history. Suddenly it's possible to operate a washing machine or garage door using a smartphone, change lock codes on doors and issue temporary lock codes for a visitor or repair technician with a phone, and assemble a home automation platform with smart lighting, thermostat, security, and other systems. Manufacturers are introducing hubs that control a growing array of devices and appliances. Major companies such as Apple, which now offers a platform called HomeKit, are in fact pushing into the home automation space.

The Rise of Protocols and Platforms

For years connected devices were mostly limited to timer boxes and hobbyist kits. Installing systems to control lights and appliances required a great deal of time and a willingness to tweak, tinker and adjust settings. Getting everything to work together was sometimes nothing short of torturous. Pico Electronics in Scotland developed one of the earliest standards, X10, in 1975. Among other things, it developed the first single-chip calculator, a programmable turntable for playing vinyl records and a remote control for lights and appliances. Later, the firm adapted the X10 interface to work with personal computers. Although the X10 protocol demonstrated what was possible using home automation, it never took off and, over the years, has drifted into relative obscurity on the home automation front. It was simply too expensive and too clunky.

But other protocols have emerged, notably Z-Wave, ZigBee, and Insteon. For instance, the Z-Wave wireless communications platform uses low-power radio frequency waves in the 900 Megahertz range to connect electronic devices, including lighting, access controls, thermostats, security devices, smoke alarms, and appliances. It is optimized for low latency and high reliability, which means it provides dependable and persistent connections between devices. In addition it exchanges small data packets at rates as high as 100 kilobytes per second. As a result it is not subject to interference from surrounding Wi-Fi and Bluetooth systems. More than 160 manufacturers now produce products using Z-Wave technology.

ZigBee provides communication protocols that create a personal area network (PAN). A PAN allows a person to connect different devices and, in some cases, uplink data to the Internet. ZigBee focuses on low-cost, low-speed communication between devices. It essentially works within a 10-meter area (approximately 33 feet) and delivers a maximum transfer rate of about 250 kilobytes per second. The platform, which uses 128-bit encryption to protect data, can transmit signals through intermediate ZigBee devices to reach more distant devices—thus creating a *mesh network* (where all devices or nodes can relay data to the network). This makes ZigBee best suited for applications that use low data rates and only intermittent transmissions. Thus ZigBee is now widely used for wireless ad hoc networks incorporating light switches, thermostats, electrical meters, health monitoring devices, and a variety of commercial and industrial systems. The ZigBee Alliance in fact boasts more than 600 products from over 400 companies.

A third major platform, Insteon, relies on radio frequency (RF) through power lines and over the air to manage light switches, lightbulbs, thermostats, motion sensors, surveillance cameras, and more. This makes it possible to bypass obstructions and interference that results from steel, concrete, and other objects that often block radio waves. Rather than adopting a command-and-control approach to managing devices, Insteon, which can transmit at a relatively fast rate of 38,400 bits per second, relies on a peer-to-peer approach. Within the network, each nonbattery device serves as a two-way repeater, meaning that the system can find the fastest available route to exchange data within the network. Insteon claims that its technology has been deployed in more than one million nodes worldwide.

Although Z-Wave, ZigBee, and Insteon are all relatively mature platforms, they're not the only protocols and platforms manufacturers use. The result is a highly fractured IoT environment that is confusing and troublesome for many consumers and businesses. The different protocols have in fact too often become separate islands—each requiring different and sometimes redundant equipment— that reduce the benefits and ratchets up the challenges of automating a home. This in turn is spawning additional systems that tie together various home automation protocols and platforms. One such product, Revolv, promises to link as many as seven different home automation technologies via a smartphone or tablet app. Just as a universal remote ties together different electronics systems, Revolv promises to connect things like lights, locks, thermostats, speakers, smart plugs, shades, and sensors under a single platform.

The IoT Meets the Real World

While some of these connected capabilities have existed in one form or another for a couple of decades, they have typically been relegated to the wealthy and technically savvy. However, a new breed of systems—typically costing hundreds or a few thousand dollars rather than hundreds of thousands of dollars—is now entering the picture. Moreover these systems are becoming smarter, cheaper, more interconnected, and better. Let's take a look at several key areas and scenarios where the IoT is fundamentally rewiring and remapping the way people go about their lives.

Home Automation Gets Real

The allure of home automation is the promise of greater convenience, improved security and greener and more efficient systems. In addition to connected lighting, garage door openers and smart locks, an array of other products are now taking shape. For example, next-generation smoke detectors can alert emergency responders when a fire breaks out. Some systems also let users silence chirps with a smartphone and notify the user when there's a need for new batteries. Meanwhile smart thermostats—besides being easy to program and adjust—optimize performance and reduce energy consumption by as much as 40 or 50 percent. Future systems will sense when a person enters a house and adjust. Over time, they will learn the living patterns of occupants and the unique characteristics of a house automatically. In the process, researchers at the

University of Virginia estimated that a typical 20 to 30 percent energy reduction would result in savings of 100 billion kilowatts and $15 billion annually in the United States alone.[3]

The technology in fact increasingly exists to control every electronic device in the home—and even manage nonelectronic devices. Smart security systems and video monitoring have already become widely available. These systems provide remote monitoring, remote arming, and disarming features, and some activate an IP camera and send a text alert when the system detects motion. Soon, these security systems will likely recognize inhabitants carrying permanent or temporary authorization tokens on a smartphones and use facial recognition to determine when someone has gained unauthorized entry. In a latter scenario the system could alert a security firm or law enforcement agency.

One of the hottest areas for the IoT and home automation is in the kitchen. Manufacturing firm LG has introduced smart appliances—including a refrigerator-freezer, washing machine, and oven—that allow a homeowner to use smartphone controls or natural language commands such as "start washing with warm water." It is also possible to change settings or start a load of laundry while away from the house. Meanwhile a smart manager built into an LG refrigerator lets a person check the contents from a smartphone via a built-in camera. The device includes a

Freshness Tracker that keeps track of expiration dates and a meal planner that makes recommendations and serves up recipes based on what's inside the fridge at any given moment.

It's not difficult to envision a day when individuals will use natural language commands, smartphone controls, and connected devices to generate shopping lists, find recipes, and accomplish other tasks. A smartphone app will guide a person to the desired products in a store. Then, once home, instead of fumbling over a byzantine collection of buttons and controls, it will be possible to pop an item in the microwave oven and say, "Defrost my frozen bagel" or "Reheat my coffee." Likewise we will instruct our television or streaming media player to switch on and tune into the content or channel we desire, in much the same way that Apple's Siri and Google Now allow users to operate their phones with voice commands today.

A Healthy Outlook

Few areas offer a more compelling argument for the Internet of Things than health and wellness. Nike Fuelbands, Fitbit wristbands, and Jawbone fitness trackers provide insights that were unimaginable only a few years ago. These devices plug into apps that share data with other apps and create an entire ecosystem of products and services in the personal fitness space—from exercise to nutrition, and beyond. There's no need to manually tabulate calories and nutritional information. These devices measure activity through accelerometers and use barcode scanners to gain a remarkably complete picture of a calories, nutrition and exercise. The data are presented to a person in charts, graphs, and images via the web or a smartphone app.

The technology is muscling into other areas as well. Connected scales beam data to servers in the cloud, which send the data to a personal dashboard on the web or in a smartphone app. Sleep systems record environmental data such as noise level, room temperature, and light and, combined with a sensor that slips under the mattress, deliver detailed information about sleep patterns and cycles throughout the night. These systems, which integrate with smartphone apps, increasingly generate personalized programs for falling asleep and waking up. In addition there are now systems for measuring and correcting posture, devices that measure exertion and oxygen consumption while working out, and pocket-size isometric trainers that provide instant feedback via a smartphone app.

Personal fitness is only one step toward a more connected future, however. A growing array of medical devices that once cost hundreds or thousands of dollars are now entering the picture. These include blood pressure monitors, blood sugar monitors, and in-home medication dispensing systems that issue reminders, dispense proper

It's not difficult to envision a day when individuals will use natural language commands, smartphone controls, and connected devices to generate shopping lists, find recipes, and accomplish other tasks. A smartphone app will guide a person to the desired products in a store. Then, once home, instead of fumbling over a byzantine collection of buttons and controls, it will be possible to pop an item in the microwave oven and say, "Defrost my frozen bagel" or "Reheat my coffee."

doses, and alert caregivers and medical professionals if something goes awry. In the not-too-distant future, doctors might also insert microsensors and nanobots inside our bodies. These devices could monitor organs and tissue and identify when a dose of medicine is required—and dispense an optimal amount. They could relay detailed information back to clinicians.

The IoT will likely revolutionize medicine. Instead of visiting a doctor once a year for a checkup that lasts only a few minutes or a nurse checking in with a high-risk patient constantly, sensors will provide continuous monitoring and data $24 \times 7 \times 365$. Using next-generation software and sophisticated algorithms, they will analyze streams of detailed data and identify potential problems and trigger points early—so that doctors and other practitioners can treat conditions in a more proactive and enlightened way.

At the same time consumers will use 3D printers to fabricate medical devices such as splints, syringes, and braces. Medical professionals will print replacement tissue such as skin and various internal organs. Indeed researchers at several universities have already successfully demonstrated so-called bioprinting. A team at Cornell University, for instance, has printed a human ear that could replace one that's damaged or cut off. Another team at the Wake Forest Institute for Regenerative Medicine is developing 3D printed blood vessels while a firm named Organovo Holdings is developing replacement livers and other organs.

Money Matters

Today people bank online, trade stocks, and other securities at the click of a button, and pay bills by clicking or tapping into an app. Mobile apps allow customers to capture a check image with the phone's camera and deposit the check without visiting the bank or an ATM. Essentially the phone becomes the bank branch. In addition smartphone-based digital wallets that allow a consumer to deposit money in a parking meter or vending machine—or pay for a purchase at a coffee house—are taking shape.

These digital payment systems are changing everything from swap meets and garage sales to retail checkouts. For example, a small device called Square plugged into an iPhone or an iPad via the audio jack creates a full-fledged point of sale device. A merchant simply slides a credit card through a slot on the top of the Square and the software completes the transaction. The system also eliminates the need for a loyalty card. A customer's visits and purchases are recorded automatically, and they're visible through a smartphone app. Not surprisingly, Square isn't the only payment processor to enter the space. Others, including PayPal and Intuit, offer similar systems.

But the Internet of Things promises to change more than banking and payments. In the insurance industry, for instance, all the innovation and data crunching from devices, sensors and systems will lead to completely different business models. The traditional approach to auto insurance relies on an aggregate model that takes into account general risk factors and expenses. However, a pay-as-you-go model has emerged. A small black box that plugs into a vehicle's diagnostic port records trip information and mileage and, using a cellular modem, transmits data to the company. The user then pays a per mile rate based on how far he or she drives—and in the future, possibly how well he or she drives, according to the little black box.

Planes, Trains, and Automobiles

Sit behind the steering wheel of a growing array of vehicles, and it's possible to see the future of the automobile—today. On board navigation systems and computers are linked to smartphones, which accommodate a growing tangle of functions, ranging from unlocking doors and starting the engine to placing phone calls and entering an address into the navigation system. These systems increasingly use voice commands—Apple's CarPlay system relies on speech recognition tool Siri, for example—to blend functionality from the phone and the Internet.

A growing number of vehicles also support mobile Wi-Fi hotspots and various features that morph driving with computing. In addition the black boxes that track trips for insurance purposes tap into the onboard computer and provide detailed diagnostics that goes far beyond status lights. A 2014 study conducted by consulting firm Capgemini found that 55 percent of current auto buyers either use connected car services now or would like to have them in the next vehicle. Only 18 percent indicated that they don't care about connected capabilities.[4]

While today's cars have many autonomous functions and advanced telematics features, including adaptive cruise control, automatic breaking, lane departure warnings, and self-parking features, fully autonomous vehicles are just around the bend. These vehicles will read traffic lights, road signs, and navigate highways and byways using sensors, satellites, and data from the Internet. Since 2010 Google has operated a self-driving automobile that uses a 64-beam laser system. The car (actually a test fleet of ten modified vehicles from Audi, Lexus, and Toyota) has, among other things, traversed San Francisco's curvy and steep Lombard Street and navigated the Golden Gate Bridge.

Further into the future, autonomous vehicles will likely navigate smart road networks and respond to traffic congestion and other issues by automatically rerouting traffic for optimal capacity and speed. These systems would also allow cars to follow closer to one another and thus increase the capacity of roads. Autonomous vehicles would also optimize fuel efficiency and reduce collisions. Studies show that upward of 90 percent of all collisions involve human error. Observers say that autonomous vehicles would result in a 30 percent improvement in fuel efficiency. They might also allow seniors to stay mobile after they are no longer able to drive on their own.

The way we think about cars could change dramatically in the years ahead. Self-driving cars could create more of a mass transit mindset rather than the current ownership model. For example, widespread car sharing could become the norm. An individual might simply order a vehicle using a smartphone and it would drive itself to the location in a matter of minutes. Once at a destination, the car will drive off to the next user.

Automated systems could also eliminate the need to park a vehicle manually. A motorist would simply step out of a car in a passenger loading/unloading zone at an airport or shopping mall and the car would park automatically and later return on command. The vehicle would use sensors in a parking garage to determine where there's an open spot. Today a number of apps have begun to deliver on the promise by finding and reserving spots in participating lots in cities such as Baltimore, Boston, Chicago, New York City, and Milwaukee. Portland International Airport uses a crude version of such as system to help motorists find open spots. A small green light appears above a stall when a spot is open and it turns red when a car is parked. Signs at the end of entry rows display which spots are open. The next step would be to connect the information to vehicle navigation systems.

But vehicles are only one piece of a connected infrastructure. Smartphone apps now provide information for subways and other forms of public transportation. For instance, in Melbourne, Australia, Yarra Trams, which operates 487 trolleys covering 29 routes and 250 kilometers of track, uses sensors embedded in tracks along with other data to provide a smartphone app, tramTRACKER, to let riders know precisely when trains will arrive at a station. The system also generates alerts when significant delays take place or a tram breaks down.

Smartphone apps are also revolutionizing everything from finding gas at the lowest price and locating a car in crowded stadium parking lot to downloading an electronic boarding pass for a flight or checking into a hotel using a barcode at a kiosk. AT&T researchers have developed a "smart luggage" tag that shows travelers a bag's location at any given moment. In addition smartphone apps now let people watch the movement of buses, trains, and airplanes in real time.

A New Era of Shopping Emerges

The Internet has revolutionized the way we search for products and shop. Phone directories have mostly disappeared, researching and buying an item such as a car or computer can be accomplished from home, and even customer service increasingly takes place online. Today, in the United States, e-commerce accounts for about 5.2 percent of total retail spending but the figure is expected to rise to about 10.3 percent, about $370 billion, by 2017. Within the United States, 60 percent of all sales will involve the Internet in some way by 2017, according to consulting firm Forrester Research.[5]

However, today, a growing number of consumers are using dedicated apps on their smartphone or tablet to buy things. These mobile tools are fundamentally altering the way purchases take place and, in the process, leveling the playing field between retailers and consumers. Cameras built into phones have become barcode readers that allow a consumer to compare prices about products on the spot. It's possible to scan, say, an espresso machine at a store and see the prices at other retailers in the area—as well as online. This practice, referred to as showrooming, has fundamentally rocked the retail industry and unleashed huge changes in the way retailers display products, provide information, and compete with online retailers on pricing and services.

Likewise apps such as Fooducate allow a shopper to scan a barcode on a product at the grocery store and view details about a food product, along with a grade. Essentially the smartphone becomes a scanning device, portable database, and meal tracker. Similar scanning apps also exist for wines, beers, and many other areas of interest.

Many of these apps also create thriving social media communities—where people share ratings, questions, and thoughts. Some merchant apps also hold electronic loyalty cards.

Not surprisingly, retailers are working to further bridge the gaps between the physical and virtual worlds. In a somewhat crude sense, Quick Response (QR) codes allow a person to scan objects incorporate them into the Internet of Things via a smartphone or other device. These two-dimensional codes, sometimes found on product packaging but also magazines and even websites, make it possible to view far more detailed information about food, housewares, electronics, and more. RFID tags and emerging technologies, such as Apple's iBeacon, take the concept to an entirely different level. They have the potential to transform shopping into a highly personalized, contextual, and interactive experience.

iBeacon, an indoor positioning system, uses Bluetooth Low Energy (also referred to as BLE or Bluetooth Smart) to communicate with smartphones and tablets in stores. When the system identifies a shopper with a compatible app running on iOS or Android device, it pinpoints the person's exact location and push messages, documents and other information to the device and receives data back. This makes it possible to send suggestions or promotions to a shopper based on what products he or she is looking at or where they're spending time in the store. A person hovering over the laundry detergent aisle, for example, might receive a coupon from a manufacturer for $1 off, if it's used on the spot.

The same technology can remind shoppers about items that are on a list, push in-store promotions or information based on a preference for, say, hats or pet supplies, direct them to pre-ordered or pre-paid items, serve up a map of seating and concessions at an arena or ballpark, deliver e-tickets, or sell discounted seat upgrades when a person arrives at a venue. Major retailers such as American Eagle, Duane Reade, Macy's, Safeway, Tesco, and Walmart have already tapped the technology in one form or another. Several major league baseball teams and the NBA's Golden State Warriors have also used iBeacon technology. Future use of beacons could include billboards that connect to automobile navigation systems and deliver targeted promotions for eateries and other establishments.

Smart shelves could further revolutionize shopping. For instance, semiconductor giant Intel Labs has developed a technology called *Shelf Edge*.[6] The prototype system connects a smartphone to Bluetooth displays in stores and allows customers to interact with smart products through the handheld device. Shelf Edge could deliver product information and even alerts based on food allergies and lifestyle preferences. Meanwhile Internet-connected coupon dispensers on store shelves could interact with customers and their smartphones on a real-time basis. Depending on the redemption rate and other factors, the manufacturer or retailer could increase or decrease the amount of the coupon or promote a different product in real time.

Accenture Technology Labs is now exploring augmented reality that could further bridge the gap between the physical shopping world and the online experience. Its *WeShop* prototype app supplements traditional product labels and information cards by providing additional data from a variety of sources. It displays social activity related to a product, loyalty program offers, buying recommendations, and other information. When a consumer holds a smartphone or tablet over a product label, the system delivers personalized information for that specific shopper. For example, if you're on a diet, the app might tell you how a product rates and offer healthier alternatives.

Others, such as Adrian David Cheok, a professor of pervasive computing at City University London, are developing devices that simulate taste, smell, and touch. This could lead to an ability to smell and taste products—from candles to restaurant items   from a computer or a smartphone. He has already built a number of devices that handle these tasks on a basic level, using chemical, electronic, and magnetic capabilities. He says that scent devices could attach to a computer and use cartridges similar to the way today's ink printers work.

The end result? Over the next few years, shoppers will see the design and layout of stores change as point of sale terminals disappear and new store layouts take shape, including salespeople handling payments with tablets or smartphones. They will also see websites incorporate new and intriguing capabilities that extend human senses and allow a person to sample a taste, smell, or feel before buying an item.

Putting Connections to Work

Over the next several years, the IoT will introduce even more dramatic changes along with entirely new ways to do things—via emerging tools such as 3D printing, digitized smell and taste, robots, and drones. These systems in turn are establishing an even broader platform for a connected world.

PUTTING THE INTERNET OF THINGS TO WORK

The IoT Meets the Real World

As far more sophisticated sensors, microchips, and data analytics capabilities take shape, there's a growing ability to observe environments and understand complex relationships. These devices—ranging from basic monitoring systems and data streams to complex biosensing devices that exist within bodies, pipes, crevices, and otherwise impossible to reach locations—are dramatically redefining how machines interact with the world around us, and how humans interact with each other.

Connecting all the digital dots is a significant challenge. Designing and building systems that truly work in the real world—and deliver maximum value—is remarkably difficult. Beyond the social and psychological ramifications, there are a number of technical and practical tripping points. Among them: interruptions to physical Internet access, component failures within systems, software bugs that generate errors and noise, data sharing across systems and organizations, coping with proprietary and competing systems, and dealing with upgrades, patches, and obsolescence.

There are also potential obstacles in terms of building out IT systems and end points that generate and capture reliable data, make it available for widespread use, and tap big data and analytics systems. Amid all of this, it's clear that governments, businesses, and individuals must approach the Internet of Things in thoughtful, multidimensional, and creative ways. It's critical to understand how people will use these systems in daily life and at work, how they might abuse or misuse systems and where the Internet of Things ultimately takes us. Will it simply speed up our already harried lives or produce genuine benefits? Will it foster mindless automation or deliver mind-bending improvements?

It's a Matter of Standards

Although much of the plumbing for the Internet of Things is already in place—ubiquitous and pervasive communications networks, sensors that can detect activities and events in the surrounding environment, and computers

Connecting all the digital dots is a significant challenge. Designing and building systems that truly work in the real world—and deliver maximum value—is remarkably difficult. Beyond the social and psychological ramifications, there are a number of technical and practical tripping points.

that can sift through massive amounts of data and transform bits and bytes into information and knowledge—society is only beginning to connect devices in any meaningful way. Just as the introduction of the web offered a crude but remarkable glimpse into an emerging virtual world, connected devices and smart systems are in the early stages of adoption. Currently they deliver limited functionality, features, and value in niche areas and specific spaces.

One of the primary roadblocks on the path to a more robust and all-encompassing Internet of Things is the battle over protocols and standards. Within the high-tech arena, this is nothing new, of course. Different hardware standards, operating systems, and file and document formats have plagued and frustrated business executives and consumers alike. Only during the span of the last few years has the computing environment matured and powerful tools and mechanisms—such as standard file formats, unified messaging and cloud computing—created bridges to a more connected digital world. This evolution, often referred to as the *consumerization of information technology*, has forwarded usability and driven gains in productivity.

At the same time there's been a march to more open standards. In 1991 Linus Torvalds released the first version of the operating system Linux, which has since mushroomed into a tech heavyweight. It's used at more than half of all corporations. More important, open systems and open source coding has sent shockwaves throughout the business world ... and beyond. The concept has fundamentally revamped the way businesses operate. From photography and personal fitness devices to industrial lighting and HVAC systems, it's becoming ever more difficult to build and sell things that work inside a walled garden. Today a product or app is merely a cog in much bigger wheel of integrated machines and code.

Breaking down the invisible walls that separate industrial systems and consumer devices is a daunting task. Many companies cling to proprietary technology because they perceive—rightly or wrongly—that it offers an advantage in the marketplace. Likewise some business executives believe that an open system or API will benefit a competitor at the expense of their company. As a result they engage in classic protectionism. Yet, at some point in the general progression of business and technology, industry and society hits a tipping point that renders a competitive advantage worthless, or worse, a competitive disadvantage.

Today, for example, even the most sophisticated and well-designed electronic typewriter or film camera offers marginal, if any, value for the masses. Moreover neither of these products represents any significant business opportunity in an age of digital devices. Likewise an app that controls functions for a specific vehicle or appliance provides marginal value. In the early stages of the Internet of Things it's enough to sell some consumers on the concept.

But expectations grow and new innovations emerge. The real gains from connected devices do not derive from using a smartphone app to start a car's engine or adjust the temperature in a house via the web. It's when vast networks of devices share data and put the data to work in ways that push past evolutionary gains and into the revolutionary category.

The challenge for businesses is navigating the boundaries between these technology states or gateways and making the transition to a new connected world. Just as proprietary networking protocols from the likes of IBM, Novell, Bay Networks, Cisco Systems, and others, eventually vanished in favor of a common standard, Internet Protocol (IP), proprietary, and closed IoT systems must eventually give way to a more open environment in order for society to realize the maximum benefits. Businesses that cling to proprietary products eventually discover they are at risk of becoming irrelevant or obsolete.

Today we take a vast array of everyday devices and systems for granted. But consider what things would be like if every automobile manufacturer used a different set of operational controls. Imagine if motorists had to use a steering wheel in one car but a joystick or control bar in another. Imagine if e-mail systems didn't connect to one another or telephones didn't work across service providers (these in fact were a problem in the early days of telephones and e-mail). Imagine if different brands of appliances required entirely different water and electrical hookups. As costs and complexity escalate, sales and adoption plummet.

Likewise, in a proprietary IoT world filled with islands of connected devices, it's next to impossible for a homeowner to manage a collection of lights, security devices, a thermostat, lock systems, a garage door, and other machines and gadgets from a central app or control panel. It's also far more challenging and expensive for a business to reach a target audience with promotions or interactive content in a mall, cinema, or sports arena—if every venue requires different apps, tools, technologies, and methods to access and process the data.

The business world is beginning to recognize the need for robust standards in the IoT space. The Institute of Electrical and Electronics Engineers Standards Association (IEEE) has established a number of standards and protocols to aid with the development of connected systems. Karen Barteleson, president of the IEEE Standards Association, describes this as the "connective tissue" for the IoT. These standards—based on an open model—encompass a number of areas, including networks, sensors, medical devices, smart homes and buildings, smart roadways, and smart city grids. Another standards group, the Internet of Things Global Standards Initiative from the International Telecommunications Union (ITU), is also attempting to build a framework of IoT standards. In addition a group called the Allseen Alliance is designing an open source platform for IoT products, systems, and services.

Businesses and government are getting into the act too. In March 2014 a group of large and powerful industrial firms, including AT&T, Cisco Systems, GE, IBM, and Intel, announced that they would cooperate on establishing engineering standards to connect sensors, objects, and large industrial machinery systems. The White House and other government entities have joined in the initiative. Together, these groups hope to take interoperability to a much deeper level. Abhi Ingle, a senior vice president of AT&T's advanced solutions group, described the current challenge in a *New York Times* article: "As an industry, we've come to the conclusion that for the Internet of Things to really take off, we needed more interoperability, better building blocks and better standards."[1]

The need for standards and protocols in fact includes everything from the way small machines use electricity and battery power to the way devices communicate and exchange data. It touches on topics as diverse as cabling technology, accounting principles, and payment systems for data operators. It also includes the way companies slot all the data into massive databases and which security standards they use. Without these common standards—and clear policies for managing data governance and other issues—the vast economic and practical potential of the IoT will never be realized.

Tackling the Adoption Curve

The Internet of Things faces barriers that extend beyond technical standards. Another formidable obstacle revolves around the money, time, and resources required to bring legacy systems, machinery, and vast industrial systems up to current technology standards. Upgrades, retrofits, or a wholesale rip and replace approach can take years—or even decades to complete. Companies often replace systems at the end of their lifespan or when there's a compelling return on investment, not when a new technology appears.

Of course, consumer preferences, technical capabilities, and business conditions change. Early adopters may realize some benefits from a connected environment—a few may achieve a commanding competitive advantage—but the first wave of adopters also assumes a greater risk of choosing a dead-end path or coping with additional costs to convert to other systems in the future. As the environment matures, prices drop and businesses and consumers prove the viability of the technology, the situation eventually reaches a tipping point and more widespread adoption takes place.

Already some companies are reaping enormous productivity gains and cost savings by switching on connected devices and systems. Global aircraft manufacturer Airbus, for example, has developed a smart factory system[2] that makes it possible to track tools, logistics media, and wing production in real time using RFID. Among other things, the system provides insight into steps, processes, and workflows that contain inefficiencies. It also makes it possible to locate tools and equipment at any given moment. The firm already tracks more than 3,000 parts per plane using passive RFID tags along with other technology.

Swedish Transport, which uses RFID and other technologies to monitor its fleet of trains over a 13,000-kilometer network of tracks, has trimmed operating and maintenance costs by at least 5 percent through connected systems. More important, notes RFID project manager Lennart Andersson, early detection dramatically reduces the risk of track damage and train derailments, which can cost hundreds of times more money and cause enormous damage to systems and people, as well as wreaking havoc with service and schedules. Finally, the technology has greatly reduced paperwork and manual tracking, which is both time-consuming and prone to errors.

Yet these types of results only scratch the surface of what's possible with connected systems. Once parts and component suppliers for Airbus and Swedish Transport—as well as a plethora of other companies across a wide swath of industries—begin integrating and embedding sensors into their equipment, the capabilities grow by leaps and bounds. Suddenly machine components and subcomponents can talk to each other and exchange information that extends deep into machinery and operational patterns. This holistic environment—with the right software—serves as the foundation for smart machines, smart factories, and even smart cities.

Building a Better Sensor

Sensors are the eyes, ears, nose, and fingers of the IoT. They are essentially the magic that allows the IoT to work. Over the last quarter century, increasingly sophisticated and increasingly tiny sensors, electronics, and nanotechnology have redefined a vast array of consumer and business systems. For example, researchers at the University of New South Wales in Australia have now developed a tiny lab-on-a-chip that can handle an array of tasks, including detecting toxic gasses, fabricating integrating circuits, and screening biological molecules.[3]

Today devices can detect and measure minute concentrations of pollution or toxic substances in the atmosphere or water supply. The can detect incredibly tiny changes in structures, such as bridges and tunnels, by measuring vibration. Sensors in vehicles allow cars to park themselves and detect when another vehicle on the road is too close. Meanwhile motion sensors in security and video systems provide alerts when an event or change takes place. This allows a person to view activity quickly and determine whether a problem exists. It also provides evidence in the event of a theft or more serious crime.

To be sure, thousands of different types of sensors now exist, incorporating light, sound, magnetic fields, motion, moisture, tactile capabilities, gravity, electrical fields, chemicals, and much more. In the past many of these sensors used analog and low-tech methods to gauge conditions for the surrounding world. For instance, for centuries people relied on thermometers with glass tubes containing mercury to measure the expansion and contraction of a liquid within a calibrated device. Likewise barometers, humidity gauges, and other devices used pressure, vacuum, and other systems to detect changes in the weather. These devices were extremely useful in the flat-earth of unconnected analog devices.

But digital technology radically changes the equation. Today's microelectronics measure many more things—and measure them far more accurately—than even the most sophisticated analog and mechanical devices of the past. They can incorporate multiple functions on a single microchip and they rely on a common binary code to transmit and receive data in real time. What's more, connecting a vast array of sensors or building them into machines, including robotic devices, provides deep insights into the interrelationships of different factors and systems in the physical world. Simply put, the technology boldly takes us where no man or woman has ever gone before.

One of the most interesting areas of sensor technology revolves around micro-electromechanical systems

(MEMS) that can be strung together into so-called mesh or smart dust networks that make them more easily incorporated into a variety of electronic components and systems. These tiny self-powered devices, in many instances measuring less than 2 by 2 millimeters (the size of a dust particle, hence their name), can be equipped with analog-to-digital converters that allow older machines to feed data to the IoT. Moreover these sensors, designed to measure everything from light and pressure to vibration and magnetism, now cost less than $1 each, compared to $10 or more a few years ago. This makes them far more cost effective for general and widespread use in fields as diverse as medicine and meteorology.

Yet today's sensors must advance considerably in order to fully tap into the Internet of Things. For example, researchers are now engineering electronic sensors that will measure smell, taste, and other sophisticated functions. This could revolutionize many industries—from food manufacturing and restaurants to detecting diseases in the early stages. The field has already moved far beyond science fiction. A July 2013 research paper[4] in the *Journal of Chromatography B* pointed out that dogs are able to identify, through smell, melanomas on humans. Using these same biomarkers, the researchers have developed a nanotechnology sensor that can sniff out the cancer in its early stages.

In addition to the research that Adrian David Cheok is conducting at City University of London—he has developed devices that communicate touch, smell and taste through the Internet—others are entering the picture. For instance, San Francisco-based Adamant Technologies is currently developing a small processor that digitizes smell and taste. It could appear in future smartphones. The system uses about 2,000 sensors to detect aromas and flavors. This compares to about 400 sensors in the human nose. The system would detect when a person has bad breath or over the legal limit to drive. The same technology could also measure a person's metabolic rate by gauging breathing and, using the IoT, tap into diagnostics capabilities to warn about an upcoming asthma attack or detect diseases such as tuberculosis and melanoma.

Other researchers are attempting to better understand the human tongue and brain and how they can build electronic systems that mimic the ability to taste. Using sensors and receptors that convert chemical substances—sugars, fats, sodium, pH levels, and other substances and properties—into tangible profiles, they're able to take machine taste to a new level of understanding. At the University of Texas at Austin, for instance, researchers have developed a sensor array that essentially serves as an electronic taste chip.[5] It can measure five categories of taste: sour, salty, bitter, sweet, and umami (the latter is generally considered a measure of appeal or deliciousness). The emerging technology has application in fields such as medicine, environment biology, chemistry, and food and beverage manufacturing.

Reliability Is Paramount

Developing near fail-safe systems—particularly in fields such as transportation and security—is critical. The convergence of different digital technologies unlocks possibilities that would have been unimaginable only a few years ago—but it also tosses out more variables, challenges, and dangers. At the heart of the Internet of Things is persistent and reliable communication between machines and people. According to securities and brokerage firm Raymond James and Associates, the number of M2M connections (excluding consumer devices) will swell from about 1.5 billion in 2012 to over 4 billion by 2017. The adoption rate will likely accelerate into the next decade and well beyond.

In order for the IoT to deliver dependable and predictable results, it's critical to build pathways that allow data to flow like a series of roads within an urban area—all while keeping critical data encrypted and secure. When one system or communications protocol isn't functioning or isn't available, vehicles—or in this case data— can detour or bypass the blockage point and continue to the destination. In some cases this means embedding multiple communications systems within devices, caching data locally on a device until a connection is available, and incorporating peer-to-peer device capabilities that allow data to flow even when an Internet connection isn't available.

Consequently system designers must build M2M systems that rely on the right combination of wired and wireless communications protocols in order to keep data flowing without interruption. On the wired side, this includes Ethernet, power line, USB, fiber optics, modem, serial cables, and LonWorks. On the wireless side, it incorporates radio frequency (RF) technologies such as near-field communications (NFC), Bluetooth, ZigBee, and Wi-Fi along with cellular and satellite systems. In some cases it will also be necessary to use multiple technologies and communications protocols within a single space or environment.

For instance, GPS won't help a motorist find a car parked in a garage because the signal cannot pass through the concrete and steel of the structure. As a result the task requires a secondary technology, such as cellular or a beacon and finder. Similarly some systems now incorporate Bluetooth technology that allows data to flow across a series of unconnected or disconnected devices—such as a smartphone or tablet—until it finds an Internet connection. At that point the data travel to the target application or database and are incorporated into the IoT.

The IoT will also require better batteries to power sensors and other devices and systems, including smartphones, tablets, and wearables. After all, a dead device is useless. One of the obvious problems today is that devices blow through batteries and demand constant recharging. However, researchers are now at work on next-generation batteries that use more advanced algorithms to know when the device is in use and when certain functions can be switched off, and different charging technologies that recharge wirelessly through magnetic induction, solar-charging layers in screens, human motion, and pulling energy from surrounding television, Wi-Fi, and cellular signals.

Putting Data into Context

As the digital age advances, the term "big data" is emerging at the center of the IoT universe. It's easy to understand why. A growing array of sensors, devices, and IT systems generate massive amounts of data. Social media, message streams, audio, and video and a rapidly expanding universe of documents add to the mix. "When you overlay a number of functions and combine everything with the right software, it's possible to create sophisticated capabilities that transcend any particular function," points out Michael Morgan, a senior analyst for mobile devices, applications and content at ABI Research. "Cameras, microphones and sensors can work together to dramatically increase the intelligence of the device. But it's critical to use the right data the right way."

What's certain is that continued advances in microelectronics will redefine the IoT and big data over the coming decades. The biggest challenges associated with incorporating sensors into the IoT isn't so much in engineering new capabilities—from breathalyzers in smartphones to RFID tags that can detect rancid food or detect minute concentrations of explosives in public places—it's building the smart systems that can accumulate the data, sort through it instantly and validate results—in the specific context of a situation.

The traditional approach, a structured database, doesn't necessarily scale well to the IoT. Even with data that contain tags and identifiers, it's tough to sort through everything and find the right data for the specific circumstances. Projects such as Open Food Facts and Simple UPC are now attempting to deliver vast databases for smartphone apps and other connected devices—though they are in the early stages of development. Unstructured data from e-mails, audio and video files, social media, and more, create even bigger obstacles.

Consequently there's a need for sophisticated algorithms and software code that can make sense of all the data. After all, it's highly undesirable for an airport to generate a bomb alert that turns out to be a false positive. Such an event could lead to panic and even injuries as officials attempt to evacuate public areas. However, an unexpected explosion would almost certainly be worse. It could unleash immeasurable damage, including widespread injuries and deaths, as well as enormous disruptions for travelers and huge economic losses for the travel industry.

Context is a key to building connected systems that work in the real world. In order to develop smart buildings, transportation infrastructure, security systems, and smart cities—each with millions or billions of objects, IP addresses, and data generation points—it's necessary to take current data management techniques to an entirely different level. When billions or trillions of devices stream data to computers—with processing taking place at various points along the way—the concept of data capture, collection, storage, and analysis changes drastically. Within such a scenario, traditional business intelligence and analytics tools simply cannot accommodate datasets of such large and complex proportions.

As the IoT grows, integrated clouds and distributed computing models will likely provide part of the solution. By tackling data processing and analytics at various points along the value chain, it's possible to scale resources and use them at the point where they're needed at any given instant. In addition highly elastic computing capacity (the ability to dial up and dial down computing resources as required through the cloud), available from a growing array of vendors, delivers a more flexible model to handle many data processing functions. In many instances they also provide easy access to low-cost open source tools that simplify the task of combining disparate data types and formats—and extract useful information about complex relationships and interrelationships.

Context is a key to building connected systems that work in the real world. In order to develop smart buildings, transportation infrastructure, security systems, and smart cities—each with millions or billions of objects, IP addresses, and data generation points—it's necessary to take current data management techniques to an entirely different level.

Yet, even with more sophisticated computing and data management models in place, the road to smart cities and other smart systems is likely to be paved with other speed bumps. Among them: questions revolving around who owns data, how organizations verify the accuracy of data, how much organizations charge to use the data, how long they can keep data, and how data are formatted for use among multiple users from a variety of industries. Consumers might also have a say in regard to data privacy. APIs and other tools that connect data raise underlying questions about ownership and interoperability.

If engineers, product designers, developers, and others, ultimately build more advanced data models and analytics systems—and there's no reason to believe they will not—the result will be remarkably advanced systems that redefine just about everything. Context-aware sensing capabilities—driven by the next generation of software and algorithms—would change the way machines operate and people view and use personal devices. For example, a smartphone could know when it is stowed in a purse or pocket—or a person is running to catch a flight—and adjust its settings, including ring tone level and do-not-disturb functions, automatically. It could also know when an individual is asleep or needs an alert to stay awake. Similarly sensors woven into clothing, shoes, and physical objects could determine—by measuring heart rate, perspiration, calorie burn, and other factors—when a runner or cyclist needs to drink water or consume an energy bar to maintain performance levels.

The same types of gains are possible in the industrial sector. Indeed they are already beginning to take shape. In Finland, for example, sensors in trash bins now send a message to a truck when a pickup is required. This has led to a 40 percent savings in waste collection costs. In Nice, France, a smart parking system alerts drivers about open spots on a real-time basis. The system has already reduced traffic congestion and CO_2 emissions. But these represent fairly crude capabilities measured against future possibilities. For example, autonomous vehicle networks would allow far more vehicles to travel on roads—in many cases caravans could travel only inches apart—with an almost infinitesimal risk of collision. But the system could also adjust traffic lights and redirect vehicle flows as conditions change. The resulting cost savings and emissions reductions would likely reach into billions or trillions of dollars a year.

In a May 2013 research paper published in *IEEE Communications Surveys and Tutorials*, Charith Perera, Arkady Zaslavsky, and Dimitros Georgakopolous[6] point out that context-aware computing takes the form of three types of interaction: personalization, which revolves around a user setting preferences and systems responding to them accordingly (e.g., programming a home automation system); passive context-awareness, where a system monitors the environment and offers appropriate options to users (e.g., receiving a coupon when entering a store); and active context-awareness, where a system continuously monitors an environment or situation and acts autonomously (e.g., if a system detects a gas leak, it automatically notifies the utility that a problem exists). Expect to see a lot more of all of these as the IoT matures and new applications stream into our lives.

The IoT: An Open Frontier

Like the Internet, the IoT will likely emerge as a quiltwork of technologies, tools, systems, and approaches unified by the common bond of connected objects and devices. The common denominator is a need for robust communications and data standards that work together and provide real-world benefits. All the various devices and systems must deliver convenience through a combination of affordability, easy setup, functionality, effective power management, a high level of flexibility and customization, integration with legacy hardware and software systems and other connected devices, and effective security and privacy protections.

We'll examine these last two issues—as well as a tangle of risks—in the next chapter. At this point the question isn't whether the Internet of Things will impact businesses and consumers; it's how big the impact will be and what direction it will take.

THE REALITY AND REPERCUSSIONS OF A CONNECTED WORLD

The Future Arrives

The history of technology is filled with optimistic, if not utopian, views of a happier, healthier, and more leisure-oriented future. However, as every new wave of technology arrives, numerous changes occur—some positive, some negative, and many entirely unintended. It's virtually impossible to anticipate where any particular technology will take society and how it will interact with a vast array of other technologies, social systems and factors.

The Internet of Things is no exception. There's little doubt that connected devices and systems will deliver far greater automation, increased convenience, and, in some circumstances, remarkable efficiency gains. The IoT also beckons with the promise of better and cheaper products and services, along with improved safety and increased human knowledge. For example, when manufacturers attach sensors to ordinary items—food packages, clothing, household appliances, and medical equipment—a very different, and potentially far better, reality emerges. It's suddenly possible to identify defects and problems and recall items quickly and efficiently.

Likewise, when a system taps into real-time data feeds and analytics to determine consumer preferences, shopping patterns, and other criteria, it's possible for a manufacturer or retailer to adjust and adapt to changes in sales or consumption dynamically—and tweak everything from sourcing and production schedules to pricing and sales to achieve optimal performance. Ultimately the ability to drill down into data redefines everything from transportation to law enforcement and agriculture to manufacturing.

Consider: a sensor-equipped sprinkler system simplifies the task of watering while introducing conservation and cost savings for a homeowner. By connecting it to the Internet, weather data could be used to adjust watering levels based on whether rain is forecast. But the same system connected across a city could further improve forecasting, water management, and utility costs. Rather than each system optimizing conditions independently, the entire network of homes and businesses operates together, and thus more efficiently. But what happens if hackers sabotage systems to stay on and drain water supplies? What is the result if terrorists hack into autonomous vehicles or cause an entire traffic grid to malfunction?

Clearly, IoT will be used in both good and bad ways. Criminals and terrorists could use commercially available drones to spy or launch attacks. The ability to hack a video camera or a device such as Google Glass and view what a person or family is doing could not only put private lives on public display, it could offer a window into confidential records or data. Suddenly, any document sitting on a kitchen counter or a desk is at risk. At the same time, what happens if a government blocks access to content through e-book readers? In a paper world, the books still exist. In an electronic world, they simply vanish. This issue bubbled to the surface in 2009, when Amazon temporarily revoked access to, ironically, George Orwell's novel *1984*, after a dispute with the publisher. Copies of the e-book suddenly disappeared from Kindle readers worldwide.

At the very least the Internet of Things will deliver new challenges and problems revolving around security, privacy, and how we go about living our digital lives. It will almost certainly create new points of contention and dispute among members of society—while raising further questions about the digital haves and the digital have-nots. What's more, the Internet of Things will require new laws along with significant and ongoing changes to our social mores.

Putting Human Factors into Motion

One of the biggest challenges with all technology is designing systems that deliver a high level of reliability and safety. While technology often removes human judgment, decision-making, and the real-world risk of inattentiveness, it also introduces new hazards and replaces the potential for smaller scale accidents and breakdowns with larger scale problems. For instance, in June 2009 a Metro subway train crashed into another train in Washington, DC. The event, which killed nine people and injured eighty, was presumably due to a computer malfunction and the operator's inability to manually apply the brakes quickly enough.

Human factors experts refer to this as the "automation paradox." As automated systems become increasingly reliable and efficient, the more likely it is that human operators will mentally "switch off" and depend on the automated system. And as the automated system becomes more complex, the odds of an accident or mishap may diminish, but the severity of a failure is often amplified. Don Norman, professor emeritus of engineering and computer science at Northwestern University, co-founder of the Neilson Norman Group, and author of *The Design of Future Things*, says: "Designers often make assumptions or act on incomplete information. They simply don't anticipate how systems will be used and how unanticipated events and consequences will occur."

Today there's no shortage of instances where humans encounter trouble with automated systems. For instance, motorists blindly follow incorrect directions provided by an automobile navigation system, even though a glance at the road would indicate an obvious error exists. In a few instances, motorists have even driven off a cliff or collided with oncoming traffic on a one-way street after following directions rather than using their eyes and minds. What's more, studies show that many motorists tend to use automation features, such as adaptive cruise control, incorrectly. In some cases, Norman says, these automated systems cause the vehicle to speed up as motorists exit a highway because there's suddenly no car in front of them. If a driver isn't paying attention, a collision may ensue.

Motorists, airplane pilots, and train operators are all prone to becoming overly reliant on automated systems—and growing more complacent about using their skills and alertness to avoid dangerous situations. Worse, designers sometimes rely on a wrong set of assumptions or an incomplete universe of facts to build a system. They may not fully understand the way people use individual devices or tools, or cultural differences. They may also overlook the way a combination of devices alter performance or behavior. In fact Norman, one of the world's leading design experts, argues that machine logic doesn't always jibe with

At the very least the Internet of Things will deliver new challenges and problems revolving around security, privacy, and how we go about living our digital lives.

It will almost certainly create new points of contention and dispute among members of society—while raising further questions about the digital haves and the digital have-nots.

the human brain. "If you look at 'human error' it almost always occurs when people are forced to think and act like machines," he warns.

The Internet of Things ratchets up the stakes substantially. Dozens, hundreds, or thousands of devices create a multitude of real-world intersection points. Moreover, with devices and algorithms communicating with each other—and different standards and quality control criteria applied by different developers and companies—there's a real world risk of building systems that do not deliver a desired level of machine-to-human communication. As Sidney W. A. Dekker, a professor in the school of humanities at Griffith University in Australia and author of the book *Behind Human Error* explains: "There is often a great deal of human intuition involved in a process or activity and that's not something a machine can easily duplicate."

The history of technology is littered with mediocre user interfaces, arcane operating controls, and performance failures. The maturation of any technology takes time, tweaking, adjusting, and fixing. Consider it no surprise, then, that home automation and connected devices have been around in one form or another for more than a quarter century. However, installing systems that actually worked in a seamless and efficient way was, for most of this period, nothing short of daunting. Many consumers who turned to X-10 home automation or the first generation or two of connected locks and lights found themselves confronted with cryptic interfaces and devices that did not work as advertised.

The Internet of Things is only beginning to reach a critical threshold of usability and practicality. As computing power has increased, mobility has advanced, cloud computing has matured, and big data and analytics have progressed, engineers, developers, and designers have begun to build connected systems that actually work. Many of these systems are now reaching a level of design sophistication that makes them plug and play. Yet the Internet of Things, particularly in the realm of the Industrial Internet, must reach a level of dependability that fosters trust. It's one thing for a single connected vehicle to malfunction. It's entirely another for an entire transportation network to fail. The latter would result in massive gridlock and widespread collisions—along with injuries, fatalities, widespread chaos, and severe economic consequences.

Yet engineering fail-safe systems for medicine, transportation, and other fields isn't impossible. Over the last quarter century, commercial airline crashes have become exceptionally rare. Redundant systems and training are critical components, of course. But the ability to use massive amounts of data—and create computer simulations and models—helps engineers better understand stresses on planes and how weather and other conditions affect structures over time. In a connected system the use of sophisticated sensors that measure vibration and other

The history of technology is littered with mediocre user interfaces, arcane operating controls, and performance failures. The maturation of any technology takes time, tweaking, adjusting, and fixing. Consider it no surprise, then, that home automation and connected devices have been around in one form or another for more than a quarter century. However, installing systems that actually worked in a seamless and efficient way was, for most of this period, nothing short of daunting.

stresses could detect metal fatigue before a serious and life-threatening issue emerges.

At the same time, IoT technologies must function on a practical level and the complexities of IoT systems must be manageable within a society. Otherwise, consumers, businesses, and governments are likely to eschew many of these systems. Janna Anderson, director of the Imagining the Internet Center and a co-author of the Pew Research Internet Project report, *The Internet of Things Will Thrive by 2025*, points out that serious potholes and roadblocks may exist.[1] "We will live in a world where many things won't work and nobody will know how to fix them," she notes.

Moreover the novelty of any new technology eventually wears thin. What was at first fresh and exciting eventually becomes mundane, even bothersome or oppressive. A good example is e-mail, which is a growing burden for many people. Today many recipients find themselves buried under piles of messages and a heavy dose of spam and malware. Similarly yesterday's leading-edge operating systems and software interfaces have become increasingly clunky and difficult to use as technology marches forward and a larger number of software applications and tools populate screens.

Ultimately IoT systems must deliver benefits for government, businesses, and consumers—without creating any clear losers. They must solve real-world problems without creating new problems or adding to existing problems, such as crime or environmental waste. But, more than anything else, connected devices must be as easy to plug in and use as a lamp or a toaster. They must deliver the right data and information in the right context at the right time—with a high level of dependability. This requires a deeper understanding of behavior and respect for security and privacy.

Smart Systems, Dumb People?

One basic concern is whether smart devices are making humans less intelligent—or changing our intelligence. Today smartphones store tens of thousands of contacts, GPS devices lead us to a destination without the need to follow a route, and wristbands with apps track our calories and fitness level in a way that nobody could have imagined only a decade ago. The downside? People cannot remember important phone numbers, using a map is a lost art, and despite unparalleled access to fitness tools, obesity and lifestyle-related diseases are a chronic problem. There's a paradox: the more things devices do for us, the less in touch we are with our natural environment and rhythms—and the less we exercise our bodies and brains.

   Psychologist and author Douglas Lisle refers to the "pleasure trap." Human brains, he says, naturally gravitate toward the simplest and most pleasing way of doing things. But the simplest way isn't always the best way.

Author Nicholas Carr, author of *The Shallows: What the Internet Is Doing to Our Brains*, questions the instant information culture of the Internet—something that is sure to accelerate with the Internet of Things. "My mind now expects to take in information the way the Net distributes it: in a swiftly moving stream of particles. Once I was a scuba diver in the sea of words. Now I zip along the surface like a guy on a Jet Ski," he wrote in a 2008 article for *The Atlantic Monthly*.[2] Although researchers are only beginning to examine cognitive thinking and how the emerging digital world shapes and reshapes it, one thing is certain: our brains are sure to adapt and evolve to accommodate the technology. Whether we gain greater intelligence or fade behind artificial intelligence remains to be seen.

IoT and the Digital Divide

When the Internet began to take shape during the 1990s, one of the biggest concerns centered on digital haves and have-nots. The so-called digital divide focuses on the potential for economic and social inequality. At a most basic level: those who have access to data, information, and knowledge are prone to benefit. Those lacking digital tools, including the Internet, are likely to further lack opportunities in education, work, and other aspects of life. As the thinking goes, the Internet magnifies these differences.

In the era of the Internet of Things, the stakes are ratcheted up further. While connected refrigerators auto-generating shopping lists or sensor-based lighting systems probably won't make or break anyone's life, technology advances could eventually leave nonconnected individuals further behind the technology curve. Some could miss out on basic tools and functions for navigating life—or they will have to work harder to get through a day or earn a decent wage. Think of it as the digital equivalent of farming with a hoe compared to a combine.

The repercussions could be significant. For example, in the health care arena, microscopic connected sensors inside the body and wearable devices on the wrist or in clothing could provide an almost unimaginable level of medical diagnostics. Physicians could identify conditions and monitor diseases in real time—and dispense medicine at an optimal level. These sensors could detect the early stages of a heart attack, stroke, or cancer and increase the odds that an individual gets help before a medical emergency occurs. Obviously those who aren't connected to these systems—and countries where the technology isn't available—won't benefit. They may have to rely on older and far less effective procedures.

A similar set of challenges exists in education. At present, schools and educators are only beginning to experiment with the IoT. But connected devices and tagged systems open the door to a slew of new capabilities, including RFID tagged research and laboratory environments, augmented reality, and far more robust learning and training capabilities using sensor-equipped tablets and other devices. Will the digitally wealthy thrive at the expense of the digitally poor? Will the ability to become digitally skilled translate into a better career? What's more, some—like Marcel Bullinga, a futurist and author[3]—have stated that the IoT could accelerate a trend toward "de-skilling." He predicts "children will learn less and achieve more."[4] There will be less need to know facts since they will be available in real time.

A Path to Downward Mobility?

One of the challenges of integrating new technologies into society is the continual displacement of jobs and workers. A comprehensive analysis conducted by *Associated Press* in 2013 found that ongoing advances in technology and increasingly connected systems are eliminating many manual tasks.[5] Jobs such as meter readers, travel agents, cashiers, and customer support representatives are on the decline. According to AP:

> Most of the jobs will never return, and millions more are likely to vanish as well, say experts who study the labor market. What's more, these jobs aren't just being lost to China and other developing countries,

and they aren't just factory work. Increasingly, jobs are disappearing in the service sector, home to two-thirds of all workers. ... They're being obliterated by technology.

Andrew McAfee, associate director of the Center for Digital Business at the MIT Sloan School of Management and co-author of the e-book *Race against the Machine, How the Digital Revolution Is Accelerating Innovation, Driving Productivity, and Irreversibly Transforming Employment and the Economy*, has noted: "I have never seen a period where computers demonstrated as many skills and abilities as they have over the past seven years." McAfee's book delivers data, examples, and research to show that advances in technology are putting enormous pressures on the average US worker and many are being left behind.

Not all technology experts and economists agree. Many point out that the same problems occurred during the transition into the industrial age during the late 1800s and early 1900s. Some amount of disruption and displacement is painful but healthy, they argue. However, it's important to recognize that the Internet of Things dwarfs past technology advances. As self-serve technologies accelerate, automation takes hold, robots advance, and nanotechnology moves from the realm of science fiction to science fact, it's apparent that society is hitting a tipping point where humans are engineering our own obsolescence in a great many areas.

It's not difficult to envision a day where restaurants and fast food establishments are filled with robots. Or shoppers grab the items they desire and leave a store without pulling out a wallet; RFID tags and an e-payment system process the transaction automatically. Or swarms of insect-sized robots construct buildings or mine for minerals deep underground. The possibilities are nearly endless—and there's the sobering possibility that machines will create a parallel intelligence that supplements or surpasses human thinking.

What is concerning about the AP report is that it found that new industries and technologies are not creating new jobs at past rates or following historical patterns. But rather, AP points out, a common refrain exists: "The developed world may face years of high middle-class unemployment, social discord, divisive politics, falling living standards and dashed hopes."

The Threat of Digital Distraction

Electronic devices such as smartphones have emerged as a hub for communication. But there's growing concern over the use of devices in automobiles, restaurants, and plenty of other situations and spaces. To be sure, they change the nature of social interaction—and many argue for the worse. Sherry Turkle, professor of the Social Studies of Science and Technology at MIT and author of *Alone Together: Why We Expect More from Technology and Less from Each Other*, believes that there's significant cause for concern. "New technologies are not coy about their aspiration to substitute relationships with technology for relationships with people," she says.

The fallout may not be entirely advantageous for humans. Studies show that attention spans are growing shorter and today's hyperlinked world is fueling an instant gratification mindset and culture. For instance, a Pew Research Internet Project found that 87 percent of teachers say that while digital tools have had a "mostly positive" impact on learning, these technologies are creating an "easily distracted generation with short attention spans."[6] Moreover 64 percent say today's digital technologies "do more to distract students than to help them academically." Other research shows that, within the workplace, many individuals spend a considerable portion of their workday browsing Facebook and Twitter.

Critical thinking skills may also be on the decline. Patricia Greenfield, UCLA distinguished professor of psychology and director of the Children's Digital Media Center in Los Angeles, found that college students who watched "CNN Headline News" with just the news anchor on screen and without the "news crawl" across the bottom of the screen remembered significantly more facts from the televised broadcast than those who watched it with the distraction of the crawling text and with additional stock market and weather information on the screen.[7] Studies in fact show that multi-tasking "prevents people from getting a deeper understanding of information," Greenfield states.

Not surprisingly, the concerns grow with automobiles and pedestrians. About one-third of all collisions occur as a result of driver distraction or inattention—often due to calling or texting. Additionally about 8 percent of pedestrian and cycling injuries in New York City between 2008 and 2011 occurred while using an electronic device such as a mobile phone or portable music player, according to one study conducted by New York's Bellevue Hospital Center.[8] The question is: Will designers and engineers design telematics systems to seamlessly integrate and manage an array of potentially complex processes or will these systems lead to further distraction?

Ironically, one potential solution—at least until fully automated vehicles are widely available—could be facial expression analysis technology and the IoT. An automobile, helm, or cockpit could be equipped with specialized cameras and sensors to detect drowsy or inattentive drivers or operators—using indicators such as blink rates, eye closure, and head motion.

Security and Privacy Concerns Grow

Over the course of the last decade, technology has generated mounting concerns about security and privacy. Data breaches are a daily occurrence and the constant spillage of private information has led to a tangle of real-world consequences, including a huge spike in identity theft. Government and corporations are at growing risk of cyberattacks and data theft. For example, a 2013 Unisys study indicates that 83 percent of Americans surveyed have a high concern about identity theft and 82 percent worry about credit card theft. Another 2013 study by global IT and security association ISACA has found that 92 percent of the public has concerns about Internet connected devices and 90 percent fear that their online data will be stolen.[9]

These are no small concerns. The Internet wasn't designed with security in mind and, in today's world, security experts are playing a game of cat and mouse with cybercrooks and hackers. As every new threat or breach occurs, security teams scramble to plug the dike. This has led to a mélange of tools, approaches, and techniques—none of which solves the problem alone. Today it's necessary to deploy firewalls, malware detection, endpoint security, encryption, password management systems, network mapping and monitoring, and much more.

Another problem is that it's next to impossible to build security into every device. Already, the IoT has encountered a number of serious security breaches that demonstrate its risks. Over the last couple of years, hackers have broken into Internet enabled baby monitors and, at least in one instance, spoken to a sleeping child. They've commandeered their way into refrigerators and television sets—sending spam and secretly taking control of these devices. Meanwhile researchers have hacked automobiles and medical devices to demonstrate vulnerabilities. The former could lead to an inability to steer or brake a vehicle. The latter could cause a defibrillator or insulin pump to fail.

Over the next few years, manufacturers and security experts must determine how to address potential IoT vulnerabilities. It will be necessary to re-evaluate security tools and approaches and apply them in different—and smarter—ways. This may necessitate device firewalls, similar to the type of firewalls used with today's computing devices and networks. This approach could limit who can access devices and data. But it might also mean excluding connected technology from certain systems. For example, California halted the proposed use of RFID in driver's licenses in 2013 because of security and privacy concerns.

To be certain, product designers and engineers face a serious dilemma. Embedding robust interfaces and controls in devices adds to convenience but also but also exposes them to attacks. However, without the ability to manage devices it might not be possible to detect an issue until the problem or hack has already presented itself and caused significant fallout. Consequently designers, engineers, and manufacturers must consider new and creative ways to approach security. They must build systems around a framework of security or risk alienating other businesses and the public.

Not surprisingly, privacy risks also grow as systems, devices, and data become more connected—and interconnected. A May 2014 report from The Executive Office of the President of the United States, *Big Data: Seizing Opportunities, Preserving Values,*[10] acknowledges that digital business and big data present formidable challenges. As the variety and velocity of data grow—increasingly fed by new data sources from sensors, machines, cameras, storage and data processing systems—several critical issues emerge, largely revolving around data personalization; data de-identification, and re-identification; and data persistence, including how it is stored and retained. "Computational capabilities now make 'finding a needle in a haystack' not only possible, but practical," the report states.

While these risks now reside somewhere between theoretical and real, they are increasingly under the societal microscope. Consider: A few years ago, retailing giant Target identified a high school girl who was pregnant based on seemingly random items she had purchased. The retailer sent promotions for maternity clothing that tipped off her unknowing father.[11] Other companies, such as banks, now use predictive analytics to identify customers likely to change financial institutions. Entertainment firms, such as Netflix, use algorithms to suggest movies, music, and other offerings.

But these capabilities pale in comparison to the type of detailed information that will be collected, managed and mined as the IoT takes shape. The same health-monitoring systems that might motivate a person to exercise and eat well could be used by an insurer to increase rates or exclude so-called high-risk patients. Likewise an employer might use a person's genome or health data—say, a genetic predisposition to heart disease, cancer, or an early indicator for a stroke—to avoid hiring or promoting an individual.

When beacons, sensors, cameras, and smart glasses become ubiquitous and the data they collect stream into a networked world, it's suddenly possible to identify where a person is and what they are doing at any given moment. Behavior and consumption patterns—for everything from food to entertainment—could become public knowledge. Moreover, as computing advances and algorithms become ever more sophisticated, systems will become better at anticipating behavior. The novelty of Target predicting when a young woman is pregnant could be eclipsed by a steady drip of analytics capabilities that determine what a person is likely to do and when that person is likely to get sick or die.

For companies building products and systems designed to operate within the framework of the IoT, there's a critical need to understand when personal data are required and when it's necessary to strip out personal identifiers. While the task seems straightforward enough, it is complicated by a growing body of digital data that makes it increasingly possible to know who someone is—even when the primary identification data are missing. For instance, a product may avoid using a static IP address, and as a result personal data may seem secure. But by plugging in information from various logs and calling records; text messages; timestamps from cell towers, tollbooths and computers; credit card transactions and other electronic records, the individual's identity is suddenly exposed.

As data stream in from drones, surveillance cameras, geolocation monitors, wearable devices, smart vehicles, smart appliances, sensors and apps in mobile phones and tablets, social media, and device logs, there's a threat of society becoming "technology frogs sitting in a slowly warming pot where we will soon be boiled by our own data," notes Rebecca Herold, an adjunct professor at Norwich University in Des Moines, Iowa. Unfortunately, she adds, lawmakers have typically addressed problems after they occur and only after "bad things have happened."

Crime and Terrorism in a Connected World

Today news headlines are rife with stories of cybercrime. Data breaches, fraud, cyberattacks, and cyber-espionage have emerged as growing menaces, threatening everything from individuals to the security of nations. Hackers and thieves breaking into industrial systems create their own set of risks and dangers—including the ability to steal data and commandeer systems. For example, in June 2010, European security officials discovered the existence of the so-called Stuxnet worm in controls used at nuclear power plants, oil pipelines, and electrical power grids. Some speculate that the sophisticated malware was designed by a well-funded private entity or a national government agency to attack Iran's industrial infrastructure, where it reportedly infected more than 30,000 machines.

Today security in many devices is extraordinarily weak. It's not a question of whether malware will proliferate, it's a question of what steps manufacturers and others will take to lock down systems adequately. In a connected world the risks grow immeasurably. While it's not possible to provide 100 percent protection and thwart all crime, the stakes are enormous—and the threats extend far beyond Internet enabled washing machines and lighting controls.

For example, 3D printers allow individuals to bypass legal controls and manufacture guns and other weapons. These plastic devices—even if capable of only a single shot—could elude metal detectors and other security devices at airports, stadiums, and other places. A group of researchers in Texas have in fact already produced and fired a 3D printed gun. A handful of organizations have stated openly that they are dedicated to defying, if not outwardly ignoring, laws about 3D weapons. Yet illicit guns are only part of the problem. 3D printing will also make it possible to produce homemade grenades and on-demand rocket launchers that could shoot down a commercial aircraft. The technology could also introduce fake merchandise and counterfeit drugs on a massive scale.

Another risk is commercially available drones. Unmanned air vehicles are now available for a few thousands dollars. Over the next few years, demand is likely to spike. The Federal Aviation Administration (FAA) in the United States estimates that 10,000 civilian drones could be flying in the United States by 2020.[12] While legitimate uses exist in areas as diverse as agriculture, mining, environmental monitoring, industrial security, weather forecasting, package delivery, and commercial photography, there's also the danger of drones being used for less noble purposes such as spying on celebrities or political leaders, stealing objects, and dropping highly targeted bombs or payloads, such as anthrax, synthetic viruses, and other bioweapons. There are also legal gray areas. For example, activist groups say they will spy on factory farms and agribusiness to monitor environmental laws and the humane treatment of animals.

Likewise swarms of insect-sized microbots and nearly invisible nanobots could handle dangerous construction and demolition tasks, find survivors after a disaster, participate in high-resolution weather and climate mapping, pollinate crops, and fight military battles. Researchers at Harvard are now engaged in the task of developing so-called robobees.[13] A number of private firms are also developing spiders, snakes, dragonflies, and butterflies that can fly, crawl, and hop into caves, cracks, crevices, and behind enemy lines.[14] These tiny devices could be equipped with a variety of sensors that exceed human senses such as sight, hearing, touch, taste, and smell.

But, like drones, the technology could be used to perpetrate a new wave of crime, murder, spying, assassinations, and terrorism. Marc Goodman, a former police officer who heads the Future Crimes Institute, a think tank and clearinghouse that studies security and risks related to emerging technologies, says that many of today's tools are nothing short of "awesome." He believes that they could "bring about great change for our world. But in the hands of suicide bombers the future can look quite different. ... We consistently underestimate what criminals and terrorists can do. ... Every time a new technology is introduced criminals are there to exploit it."[15]

A New Legal Framework Emerges

The Internet and digital technology have ushered in massive changes to legal systems worldwide. There's a growing and often contentious focus on rights, responsibilities, and resources in areas as diverse as intellectual property, copyright and trademarks, defamation, crime, and cyberspying. Jonathan Bick, an adjunct professor of Internet Law at Rutgers University Law School, explains: "The legal system is struggling to keep up with today's technology." A fundamental issue, he notes, is that there's really no such thing as international law. "There are bilateral treaties, conventions and agreements in place that attempt to create order. But these laws are only as good as they are enforced."

One of the biggest obstacles: "What's illegal in one country may not be illegal in another," says Pauline C. Reich, director of the Asia-Pacific Cyberlaw, Cybercrime and Internet Security Institute in Tokyo, Japan and co-author of *Law, Policy and Technology: Cyberterrorism, Information Warfare and Internet Immobilization*. In the end, this makes it difficult to sort out issues relating to jurisdictions and enforcement. The challenges and problems grow exponentially as data move across servers, clouds, and devices. Attempting to understand where data reside and who has claim over data is a next to impossible task. Many say that modern computers and communications have pushed the law far beyond what it was ever intended to address.

The Internet of Things promises to heap additional layers of complexity onto an already complex environment. Attempting to understand where data originate, how data are altered or changed along an electronic pathway, creates huge challenges. In fact, as more and more houses and businesses get connected, a few key questions arise: Who exactly is responsible for a problem, breakdown, or outage, particularly if it results in damage, injuries, or deaths? What happens when a country or jurisdiction won't cooperate with the rest of the international community? And what happens when deeply personal or private information goes public due to a series of unfortunate events— none of which is specifically the cause?

Additionally there are practical and regulatory issues to sort out, besides a need to examine everything from online contracts and user agreements to surveillance and the extent of privacy protections. Perhaps only one thing is certain: the years ahead present enormous technical and practical challenges. An a globalized and interconnected IoT world takes shape, society and the legal system will be hard-pressed to manage a technology framework that is advancing rapidly and changing so many things in such profound ways. The ultimate challenge is balancing risks and protections with basic rights and freedoms.

Future Tense

The future isn't without hope. Goodman and other experts say that one highly effective approach may be to use crowdsourcing techniques to better understand how to build protections for this new and emerging connected world.

He points out that the Organized Crime and Corruption Reporting Project now tracks the activities and spending patterns for suspected cybercrooks and terrorists around the world. He believes that tossing out problems and challenges to the public—essentially turning security into an open source project—could provide enormous benefits.

But, at the same time, governments, businesses, educational institutions, researchers, and ordinary citizens must examine and reexamine consumption, convenience, and personal boundaries at a level never before considered. There will be a need to rethink and remap laws, social mores, and basic approaches to security and privacy. There will be a need to build new security tools and think in new and creative ways. Only then can society realize the full potential of connected devices, systems, and technology.

A NETWORKED FUTURE EMERGES

A New Frontier of Technology Takes Shape

As the Internet of Things and connected devices become part of our lives, a remarkable future is taking shape. Today human error accounts for 70 to 80 percent of vehicle collisions, according to the US Department of Transportation. The World Health Organization reports that 1.24 million road traffic deaths occur each year. Autonomous vehicles could virtually eliminate injuries and deaths. Self-driving cars operated within a vast network of synchronized traffic signals and routing systems could also usher in costs savings related to operating vehicles more efficiently and better maintaining infrastructure.

In the health care and wellness arena, the Internet of Things will revolutionize healthcare and telemedicine. It will introduce 24 × 7 medical monitoring and tap into 3D printing to generate medical devices and replacement organs. Tiny devices will release medication in the exact dose that's needed and exactly where it is needed—reducing side effects and boosting efficacy. These systems—along with increasingly sophisticated fitness bands and food and sleep monitors—will introduce a more tangible way for individuals to track their health and wellness. The US Centers for Disease Control estimates that type 2 diabetes could affect one in three Americans by 2050. Today, 1 in 4 deaths in the United States result from heart disease. Most of these deaths are entirely preventable through better diet and exercise.

Within industry, connected machines will allow manufacturers to provide real-time updates about order status and supply chain requirements while farmers will use sensors and other devices to optimize watering and soil conditions. They will apply pesticides and fertilizers at highly targeted and optimal levels. Meanwhile robots and drones—including insect-sized microdrones—will assist in manufacturing, trash collection, snuffing out fires, defusing bombs, and handling other tasks. In November 2013 retailing giant Amazon.com announced plans to deliver packages via drones within the next few years. FedEx chairman Fred Smith stated in 2009 that the future of package delivery lies in fleets of drones. (For now, those plans will have to stay on hold within the United States. In June 2014 the Federal Aviation Administration banned the use of commercial drones for package delivery for the immediate future.[1])

The Internet of Things isn't just about locating objects and using them to sense the surrounding environment— or accomplish automated tasks. It's a way to monitor, measure, and understand the perpetual motion of the world and the things we do. The ability to peer into the spaces between objects, people, and other things, is just as profound as the objects themselves. The data generated by the IoT will provide deep insights into physical relationships, human behavior, and even the physics of our planet and universe. Real-time monitoring of machinery, people, and the environment creates a model for reacting to changing conditions and relationships—faster, better, and smarter. McKinsey Global Institute estimates that the economic impact of the IoT will range between $14 trillion and $33 trillion a year in 2025.[2]

A number of researchers and a handful of companies are now taking the concept of a connected world to a whole new level. These concepts sound like something straight out of a science fiction novel. For instance, a *Slate* magazine article titled, "Google's Eyes in the Sky," posits that the company's foray into drones, satellites, and balloons is at least partly about creating mechanisms that can index and track the physical world in a way that approximates how Google now spiders and tracks the virtual world.[3] With cameras and various sensors in the sky and around the planet, new and remarkable windows open to data. Suddenly it's possible to watch planes, trains, automobiles,

The Internet of Things isn't just about locating objects and using them to sense the surrounding environment—or accomplish automated tasks. It's a way to monitor, measure, and understand the perpetual motion of the world and the things we do. The ability to peer into the spaces between objects, people, and other things is just as profound as the objects themselves. The data generated by the IoT will provide deep insights into physical relationships, human behavior, and even the physics of our planet and universe.

and pedestrians move in real time. It's possible to understand patterns and relationships in a way that blows the doors off today's systems. At some point in the future, the article points out, it might be possible to estimate changes in a country's GDP on a daily basis.

Amid all the possibilities, one fact stands out: the Internet of Things will revolutionize both developing and developed nations and introduce a tidal wave of commercial and consumer applications—from smarter utility grids and smart cars to radically different health care and manufacturing systems. It will change our perspective of the world and usher in automation and entirely new ways of interacting with the world around us. Along the journey, our lives will change immeasurably. While many of these capabilities may seem futuristic and even far-fetched, the next quarter century will usher in changes that are truly mind-boggling.

Let's examine how a connected future might look ...

Forward Thinking

In March 2014 the Pew Research Center released a report based on a comprehensive study of the web and Internet. *Digital Life in 2025*[4] tapped the expertise of more than 2,500 technology experts to paint a picture of where society and life are headed over the next decade and beyond. Naturally a wide range of opinions surfaced. Some observers foresee a utopian future while others expressed concern about a decidedly dystopian existence. A wide range of thoughts, opinions, and predictions covered everything from robots and 3D printing to augmented reality and highly connected and automated environments. Pew also canvassed a cross section of interaction points, including health, education, work, politics, economics, and entertainment.

Amid all the predictions, a few comments stand out. The vast majority of participants believe that the Internet of Things will lead to global, immersive, invisible, ambient networked computing environment that relies on smart sensors, cameras, software, databases, and massive data centers. Within this space, augmented reality will transpose real world input with virtual data and images displayed on wearable or implanted technologies. There will be massive tagging of physical objects to capture data that once escaped our senses. And all of this will lead to enormous disruption—socially, politically, and across a wide swath of industries, education, and government.

Some, like David Clark, a senior research scientist at MIT's Computer Science and Artificial Intelligence Laboratory, noted:

> Devices will more and more have their own patterns
> of communication, their own "social networks,"
> which they use to share and aggregate information,

and undertake automatic control and activation. More and more, humans will be in a world in which decisions are being made by an active set of cooperating devices. The Internet (and computer-mediated communication in general) will become more pervasive but less explicit and visible. It will, to some extent, blend into the background of all we do.

Daren C. Brabham, a professor at the Annenberg School for Communication and Journalism at the University of Southern California, predicted:

We will grow accustomed to seeing the world through multiple data layers. This will change a lot of social practices, such as dating, job interviewing and professional networking, and gaming, as well as policing and espionage.

Nishant Shah, visiting professor at the Centre for Digital Cultures at Leuphana University in Germany, observed:

It is going to systemically change our understandings of being human, being social, and being political. It is not merely a tool of enforcing existing systems; it is a structural change in the systems that we are used to. And this means that we are truly going through a paradigm shift—which is celebratory for what it brings, but it also produces great precariousness because existing structures lose meaning and valence, and hence, a new world order needs to be produced in order to accommodate for these new modes of being and operation. The greatest impact of the Internet is what we are already witnessing, but it is going to accelerate.

Robert Cannon, an Internet law and policy expert, said:

The Internet, automation, and robotics will disrupt the economy as we know it. How will we provide for the humans who can no longer earn money through labor? The opportunities are simply tremendous. Information, the ability to understand that information, and the ability to act on that information will be available ubiquitously. … Or we could become a "brave new world" where the government (or corporate power) knows everything about everyone everywhere and every move can be foreseen, and society is taken over by the elite with control of the technology. … The good news is that the technology that promises to turn our world on its head is also the technology with which we can build our new world. It offers an unbridled ability to collaborate, share, and interact. "The best way to predict the future is to invent it." It is a very good time to start inventing the future.

Within this emerging IoT framework, a dizzying array of issues, questions, and challenges arise. One of the biggest questions revolves around living in a world where almost everything is monitored, recorded, and analyzed. While this has huge privacy implications, it also influences politics, social structures, and laws. Jonathan Grudin, principal researcher for Microsoft Research, believes that the impact of making so much activity visible is that it will expose the gap "between the way we think people behave, the way we think they ought to behave, the laws and regulations and policies and processes and conventions we have developed to guide behavior—and the way they really behave," he explained in the Pew report. Adapting and adjust to this altered state of reality will be no easy task.

Grudin points out that society often formulates rules knowing that they won't always apply, and ignores many inconsequential violations. But in a highly connected world, that may no longer be possible. "The violations are visible, selective enforcement is visible, yet formulating more nuanced rules would leave us with little time to do anything else." What's more, data and information gathered digitally, without the full context, can be deceptive and extraordinarily misleading. "Human beings are flexible, yet we have some fundamental social and emotional

Within this emerging IoT framework, a dizzying array of issues, questions, and challenges arise. One of the biggest questions revolves around living in a world where almost everything is monitored, recorded, and analyzed. While this has huge privacy implications, it also influences politics, social structures, and laws.

responses; how technology will affect these must be worked out," he noted.

Author and MIT professor Sherry Turkle says that the intersection of technology and human interaction will play out in other ways, including how we raise children, how we deal with the aged and elderly, and how we form relationships. In a 2011 interview I conducted with her she stated:

> When we even contemplate giving the care of children to a robot, we embark on a "forbidden experiment." The healthy development of a child depends on being exposed to the full range of human expressions and vocal inflections. It depends on that child feeling love and care of a person who knows how to love and care. None of this is available from a robot. And our elders—and one day we will all be our elders—want to talk about the meaning of their lives with those who understand what a life is. And what things have human meaning—the recollection of a child's birthday, of a marriage, of the loss of a spouse. Robots can understand none of this.

Turkle identifies an ongoing societal theme of neglecting the underlying problem and, instead, engineering technology in a misdirected attempt to solve and resolve the symptoms. For example, "When people talk to me about their fantasies about robots, they talk about how people have disappointed them. I don't see robots as a solution—for robots cannot give us the love and care we need and deserve. I see our preoccupation with 'caring machines' as a symptom for how we have disappointed each other. … The fantasy is that we will somehow 'offload' or 'outsource' those things that we are finding hard to keep up with as a society." In the end, the irony is that we embrace technology such as the Internet and the IoT to make our lives simpler and easier but the result is exactly the opposite. "We turn to technology to help us find time. But we end up spending more time with technology and less with each other. And there, again, is the vicious circle."

Make no mistake, finding ways to connect and interact in a human way will emerge as a challenge in the next quarter century and beyond. A growing body of research shows that rising levels of depression and dissatisfaction in society can at least partially be attributed to diminished human contact and connections. As we embrace more technology and a greater number and array of connected and automated systems, the challenge will be to balance our desire for new and better things with our basic emotional and practical needs. After all, in the end, no matter how many devices and machines we have and how connected we are—we are human beings. It's doubtful whether any robot or system, at least in the foreseeable future, will match human complexity and thinking.

2025: A Day in the Life

Although the IoT is only beginning to take shape, it's clear that it will have a profound impact on life and business. Here's a brief look at how the day of a typical family might take place in just over a decade:

At 7 am on a Monday morning, Mary Smith wakes up after her pajamas send a mild sensory alert to her skin. A few minutes later, she climbs out of bed, steps into the shower that's equipped with a sensor that automatically switches on the water to the exact temperature she desires. The shower is connected to a smart water heater that knows the family's bathing patterns and adjusts the temperature accordingly. It can also be programmed to go into a vacation mode using a smartphone. As Mary walks from one room to another, lights turn on and off automatically. The combination of motion sensors and a software beacon on her smartphone and in her clothing react to her presence and anticipate her movements. Old-fashioned light switches are also available, if she needs to use them.

After getting dressed, Mary heads downstairs, where the coffee machine has already brewed a hot latte, based on the time she left the shower. Mary grabs a container of yogurt from the refrigerator, which adds the item to her shopping list. After breakfast, she says goodbye to her husband, John, and her two children, James and Michael. He will make sure the boys get off to school and then heads into his home office. John grabs a frozen bagel and pops it in the microwave. He taps a bagel icon on his smartphone app and the microwave oven defrosts it. He then pops it in the toaster and, once it's done, spreads on the cream cheese.

As a marketing executive for a large consumer products company, John works at home most days. The computer recognizes when he is nearby; it uses biometric authentication to log him in, and immediately displays a dashboard of key metrics. Messages and files he worked on from a tablet the previous night appear in the exact state he left them on that machine. All the data are synced through the cloud. John only occasionally heads out for appointments. When he does, as he will today, he uses a shared vehicle. He subscribes to a service that delivers an available car within ten minutes. He pays for the car based on time and mileage.

When Mary, a physician, heads to work at a local hospital, she checks her smartphone before leaving to catch any traffic alerts and view her schedule. With this information in hand, she can tap on a preferred route or let the vehicle automatically adjust the route to navigate the current traffic conditions. Since she has an autonomous vehicle that requires no intervention on the 12-mile commute to work, she typically listens to the car computer system read her e-mail and text messages and she dictates responses. When she is finished, she issues a voice command for music or other media. Along the way a billboard for a local

restaurant (she has opted into) sends a coupon for breakfast to her navigation system, but she chooses to ignore it.

At the hospital Mary leaves her car at a curbside check-
in area. The car parks itself in an adjacent structure. She
steps through the entrance and her RFID-enabled badge
lets the receptionist know that it is time to begin prepping
the first patient. Once in her office, Mary grabs a tablet
computer that displays the patient's chart—along with a
dashboard showing scores for vitals, diet, fitness, compli-
ance, and other areas. The data—including blood sugar
level, heart rate, blood pressure, cholesterol level, and body
temperature—stream in from sensors embedded in clothing and a smartwatch/health monitor worn on the patient's
wrist. Mary can tap on any topic on the screen to drill down
and view far more detailed information—including live epi-
demiological data tracking a current flu outbreak.

The hospital is quite different from the facilities we
know today. Patients wear RFID bands on their wrist to
track their movements and ensure that they are receiving
the right medicines and meals. If a nurse begins adminis-
tering the wrong medication, the system generates a visual and auditory alert. Patients use their smartphones or
hospital tablets to request nurses, change channels on the
television, and order meals. Nurses, therapists, and technicians view medical data and information—including X-
rays, ultrasounds, and pharmaceutical information—on
tablets as well.

It's virtually impossible to find a piece of paper anywhere in the facility. Doctors transfer files and information to pharmacies and patients electronically. What's more, it's possible to locate the nearest device or medical equipment because everything is tagged with RFID and visible on the network. It's also possible to track blood supplies and other essential items in real time and anticipate when there's a need to place an order.

All the data are fed into the hospital's database, where the data are analyzed to better understand patient demand, usage patterns, and treatment modalities. The analytics system digests an ongoing feed of data to determine when and how to treat patients and, based on patient, hospital, and societal variables, what approach or therapy to use.

At lunch, Mary taps an order on her smartphone, heads to the cafeteria and picks up her meal. A digital wallet in her phone automatically pays for the food. While at lunch, she receives a message from her nephew, Austin, who is visiting town for a few days. Mary issues a digital key for the house through her phone. When Austin arrives from the airport, he simply taps his smartphone to gain entry to the house. In three days, when he leaves, the key will be revoked.

John's marketing job has also changed considerably over the last decade. From his home computer, he is able to view a real-time feed of marketing and sales metrics. He can watch the computer automatically adjust and adapt pricing based on live sales and product availability at any given store location. When John receives an alert or notices a sales decline, he can order the system to send out a coupon—but only to those who are in stores or likely to enter the store based on their shopping list (which is generated by sensors in cabinets and the pantry). John can also view aggregate data through the company's loyalty program and generate a promotion. Finally, because every item is tagged, he can find the specific product if a recall takes place.

Back at home, a fleet of small robotic devices make the beds, tidy the rooms, clean the counters, vacuum the floors, and water plants inside the house. They also serve as a security system, monitoring for unknown and unauthorized intruders. Each is equipped with a camera and audio sensor to provide John and Mary eyes and ears when they are not around.

At the end of the workday, Mary heads home. She stops at the grocery store to buy a few items. A smart shopping cart connects with her phone and displays her shopping list. The system directs her to the items she desires, though, along the way, she also receives promotions for other products that manufacturers want her to try. As Mary pulls items from shelves, she drops them into a reusable grocery bag. When she has everything she needs, she simply leaves the grocery store. Because all the items are tagged with RFID, a reading device tallies the total and processes the electronic payment. She immediately receives an e-receipt in her e-mail.

Later, she and John prepare a meal that the computer has suggested—based on what's in the refrigerator and in the pantry. The system learns their preferences over time but also incorporates wellness data from wristbands, clothing, and other sources to adjust and optimize nutrition and calories.

After dinner, they help James and Michael complete homework on tablet devices, and then decide to watch a documentary about the future of technology with their children. They learn about new systems that will in only a few short years create smart grids and smart cities that adjust dynamically to constantly changing traffic patterns, usage patterns, weather, and a number of other variables. These systems promise to drastically reduce collisions, cut energy usage, and save time and money for everyone. Later, the boys use connected gloves and goggles to travel to a virtual zoo, where they can feed a giraffe and feel its tongue and pet a lion. The sensation is completely authentic.

Later, after the children are sleep, John checks out a new restaurant by sampling the flavors using a small device that connects to his computer. He books a reservation for Friday night. Afterward, he views an alert that a rain gutter is clogged. The system has automatically notified a company to come out and clear it at a previously agreed-upon price. The couple read e-books and, after a few minutes, switch off the lights using a voice command. Sensors in the bed and in their clothing track their sleep patterns, and when morning comes, the system begins to slowly adjust the lighting to aid them in waking up.

Obviously this scenario doesn't encompass every possible aspect of the Internet of Things. John and Mary are likely to encounter numerous other systems during the course of a day and into the evening. They might also have to deal with annoyances and problems, such as privacy settings and potential fraud within their e-payment system. But their lives will certainly be far different than ours today. Technology systems will be more deeply and broadly woven into their daily existence.

Left to Our Devices

Although it's impossible to predict exactly how the future will unfold—history has taught us that lesson quite eloquently—it's clear that the IoT isn't just a passing fad or gimmick. The ongoing advance of technology and the overlap of a number of fields—mobility, robotics, sensors, augmented reality, analytics, artificial intelligence, M2M communication, and more—will accelerate the march toward a connected and interconnected world. New products, services, and capabilities that we cannot yet imagine will be spawned as a result of the Internet of Things.

In a best-case scenario the Internet of Things will enrich and improve our lives in remarkable ways. Connected devices and machine intelligence will automate scores of rote functions—from shopping lists to watering a garden. They will help us live healthier—sleep better, manage our weight and exercise, and receive medical attention when we need it and in the form we need it. Monitors woven into smart clothing or positioned inside the body will detect an impending heart attack or stroke before it occurs and help doctors take action proactively rather than after the damage is done. Connected devices will also lead to safer and better vehicles, industrial machines, and perhaps even an ability to predict earthquakes, floods, and other events. Likewise far more energy-efficient and environmentally sound practices will take shape in our homes and businesses.

In a more utopian vision of the future, smart machines will continually learn and refine their algorithms and coding in order to engineer solutions to the same problems the technology creates. For example, if a hacker breaches a system, it might detect the anomaly and go out to the connected world to find the data it needs to engineer a solution. Once the device succeeds at this task, it adapts its programming to block future hacks and attacks and shares the solution with other connected machines. Within this new machine order, robots and other devices might also gain an emotional spectrum that approximates—or at some point—exceeds human capacity.

According to the Pew Survey, 83 percent of the experts it polled—from industry, academe, consulting, and law—believe that the IoT will have "widespread and beneficial effects" by 2025. But it's also very possible that the group Pew interviewed is overly optimistic. Many live and work in industries and fields that focus on exploring and developing new technologies and, in some cases, benefit by marketing and selling them. Although many identified caveats and concerns, there's also a distinct possibility that the IoT will unveil a far more dystopian future that approximates George Orwell's *1984*. This might include technology that's unmanageable and uncontrollable, systems that lead to an enormous uptick in cybercrime and cyberwarfare, a world without any sense of privacy, and greater political and social discord.

In reality, the IoT will likely land somewhere between the two ends of the spectrum. It will introduce plenty of frivolous and useless devices that quickly disappear but also deliver highly practical systems and solutions that improve the quality of life. It will make things easier and safer in some ways but more difficult and challenging in others. Like the transition to the industrial age, it will displace workers and make some jobs obsolete while introducing new high-skill careers. A connected world will generate enormous stress for some members of society—particularly those who are older—but energize and excite others. Like the introduction of any technology—the printing press, the cotton gin, the telephone, the automobile, and the computer—there will be countless winners and losers.

Only time will eventually reveal these answers and let us know if a connected world really equals a better world.

GLOSSARY

3D printing
These devices manufacture physical objects using computer software. They frequently rely on computer-aided design (CAD) software to produce functional three-dimensional objects.

Application programming interface (API)
These software programs incorporate protocols, tools and other resources used by developers to build interoperability across programs running in the same environment.

ARPAnet
An early packet-switching network that served as the foundation for the modern Internet. The network was initially funded by the US Advanced Research Projects Agency (ARPA), the precursor to today's Defense Advanced Research Projects Agency (DARPA).

Algorithm
A highly structure set of instructions or specific procedure designed to perform a specific operation or task within a finite set of steps

Artificial intelligence
Software that uses algorithms and a highly complex rules-based structure to boost computational decision-making that approximates or exceeds human capabilities.

Augmented reality
The use of technology to enhance reality by displaying a text or pictorial overlay on an image displayed on a smartphone, smart glasses, or other device.

Autonomous vehicles
A computer-controlled car equipped with sensors, computers, and other technology that allows the robot vehicle to drive itself—with no human assistance.

Big data
The use of large and broad data sets along with analytics to understand events, trends and activities in much deeper and useful ways.

Bluetooth
An open standard for wireless digital communication over short distances—roughly 10 meters. The radio frequency technology allows devices to transfer digital audio, video, text, and signals, such as from a wireless keyboard to a tablet device.

Byte
A standard unit of measure for computing. A byte is comprised of eight binary digits incorporating alphanumeric characters. Storage systems include bytes, kilobytes, megabytes, gigabytes, terabytes, petabytes, and exabytes.

Cloud computing
The use of remote servers, storage devices, and other computational tools to provide services, including Software-as-a-Service and Infrastructure-as-a-Service.

Connected devices
Various industrial machines and personal devices that connect to one another through a network, such as the Internet.

Contextual awareness
The ability of a machine or device to recognize environmental factors, user behavior, and other data to determine how to operate in a given situation or moment. For example, a smartphone might adjust the microphone or lighting to fit the noise or illumination in a particular situation.

Cybersecurity
Security relating to online systems and devices, including connected devices. This can include physical tools, hardware, and software systems.

Encryption
The scrambling of sensitive data or information so that it is unreadable to anyone other than the sender and the intended recipient. Encryption software

uses a series of mathematical formulas to scramble and unscramble text and other data.

Ethernet
A group of computer networking protocols that allows data to travel over cables, usually over a local area network (LAN).

Geolocation
The use of specific coordinates to identify the specific location of an object. Geolocation uses satellites, cellular technology, Wi-Fi, and other systems to provide specific or general information.

Global positioning system (GPS)
A system that uses satellites in space to pinpoint the location of objects on the earth's surface, including vehicles, smartphones, and other computing devices.

Human-to-machine communication (H2M)
The interaction between humans and computing devices, typically using keyboards, a mouse, a touch screen, and speech controls.

Industrial Internet
A term coined by GE to describe the use of connected machines, software, data, and analytics and wireless technology to introduce communication among machines and humans.

Internet
An infrastructure used to connect computers and other devices to each other over a common network. Today the Internet is a global network that uses TCP/IP protocol and domain name system (DNS) to provide each device with a unique address.

Internet of Everything (IoE)
A term coined by Cisco Systems to describe the sum of all connected systems, including the Internet of Things.

Internet Protocol (IP)
A communication protocol used as a networking standard for the Internet. It allows computers to handle packet switching, routing, addressing, and other functions.

Local area network (LAN)
A group of connected devices, including computers and peripherals such as scanners and printers, that communicates in real time over cable or wireless systems using a common protocol.

Machine-to-machine communication (M2M)
The ability for computing devices and other machines to exchange information and perform actions using software—without the involvement of humans.

Nanotechnology
Systems that manipulate and manage processes on atomic, molecular, and supermolecular scales.

Near-field communication (NFC)
A wireless communications technology that allows objects with NFC and various computing devices to exchange data with little or no human intervention.

Personal digital assistant (PDA)
A handheld computing device that allows a user to enter text, drawings, and other data, including through a camera or barcode reader. In the past this included devices such as the Palm, which have been rendered largely obsolete due to the widespread adoption of smartphones.

Personal area network (PAN)
An interconnection of various devices used by a single person within a restricted area, typically about 10 meters.

Radio frequency identification (RFID)
A wireless technology that uses either passive (nonpowered) or active (powered) tags (integrated circuits) and readers with antennas to identify objects

and transmit data about their condition or position to computers. RFID tags carry data that range from simple information to complex instructions.

Real-time location systems (RTLS)
A system that uses radio frequency (RF) tags to automate tracking on a continuous basis. By contract, RFID tags are read only as they pass by a fixed point with a reader.

Robotics
The branch of computer science and engineering that revolves around developing machines capable of high-precision tasks. Robots increasingly use AI to operate and interact with changes in the surrounding environment.

Sensor
A device that detects changes and variations in the surrounding environment. Sensors are increasingly added or able to communicate with smartphones and other computers.

Smartphone
A mobile phone that incorporates sophisticated sensors and a variety of digital computing capabilities, including a camera, GPS, and electronic data exchange.

Tablet computer
A multimedia computing device, such as an Apple iPad, that features an LCD touch screen and Internet connectivity through wireless, cellular, or both.

Telemetry
The ability of machines to communicate with one another (M2M) and exchange data with computers and other systems through advanced telecommunications features.

Unmanned air vehicles (UAV)
An aerial vehicle that operates without an onboard crew; usually a human flies the plane remotely. These machines, typically referred to as drones, are now used for combat as well as a growing number of business scenarios.

Wearable computing
The use of wearable computing devices, including glasses or goggles, clothing, wrist bands and watches, shoes, and other items. These objects use built-in sensors and communications systems to exchange data with smartphones and other computers.

NOTES

Chapter 1

1. Pew Internet and American Life Project, August 2013.

2. Cisco Systems, How Many Internet Connections are in the World? Right. Now. http://blogs.cisco.com/news/cisco-connections-counter.

3. Internet of Things vs. Internet of Everything: What's the Difference? ABI Research, May 2014, p. 2.

4. Full text available at: http://www.rfidjournal.com/articles/view?4986.

5. Identity Theft Resource Center. ITRC 2013 Breach List Tops 600 in 2013. http://www.idtheftcenter.org/ITRC-Surveys-Studies/2013-data-breaches. html.

6. The Evolving Internet: Driving Forces, Uncertainties, and Four Scenarios for 2025, 2010, http://newsroom.cisco.com/dlls/2010/ekits/Evolving_Internet_GBN_Cisco_2010_Aug_rev2.pdf.

Chapter 2

1. http://www.mckinsey.com/features/sizing_the_internet_economy.

2. BCG Report, The Connected World, The Internet Economy in the G-20: The $4.2 Trillion Growth Opportunity. The Boston Consulting Group. March 2012. http://www.bcg.com/documents/file100409.pdf.

3. http://www.corp.att.com/attlabs/reputation/timeline/46mobile.html.

4. June 2006 issue of *Wired*: http://archive.wired.com/wired/archive/14.06/crowds.html.

5. Cisco Visual Networking Index: Global Mobile Data Traffic Forecast Update, 2013–2018. http://www.cisco.com/c/en/us/solutions/collateral/service-provider/visual-networking-index-vni/white_paper_c11-520862.html.

6. http://www.siemens.com/press/pool/de/feature/2014/corporate/heuring factsheet-en.pdf.

Chapter 3

1.http://www.mckinsey.com/insights/high_tech_telecoms_internet/the_internet_of_things.

2. Wipro. *Big Data. Catalyzing Performance in Manufacturing.* 2011. http://www.wipro.com/documents/Big%20Data.pdf.

3. McKinsey and Company, McKinsey Global Institute, *Big Data: The Next Frontier for Innovation, Competition and Productivity.* June 2011. http://www.mckinsey.com/insights/business_technology/big_data_the_next_frontier_for_innovation.

4. McKinsey and Company, *McKinsey Quarterly*, the Internet of Things. March 2010. http://www.mckinsey.com/insights/high_tech_telecoms_internet/the_internet_of_things.

5. American Security Project. *The US and Its UAVs: A Cost–Benefit Analysis*, July 24, 2102. https://www.americansecurityproject.org/the-us-and-its-uavs-a-cost-benefit-analysis/.

Chapter 4

1. Organization for Economic Co-operation and Development. http://www.oecd-ilibrary.org/docserver/download/5k9gsh2gp043.pdf?expires=1403305175&id=id&accname=guest&checksum=EBFBAB32465D093454D55C4FB4288A20.

2. NPD Group. March 31, 2014. https://www.npd.com/wps/portal/npd/us/news/press-releases/mobile-devices-help-boost-home-automation-usage-and-awareness-according-to-the-npd-group/.

3. The Smart Thermostat: Using Occupancy Sensors to Save Energy in Homes. http://www.cs.virginia.edu/~whitehouse/research/buildingEnergy/sensys10thermostat.pdf.

4. *Cars Online 2014*, Capgemini Consulting. http://www.capgemini.com/cars-online-2014#report.

5. Forrester Research, *U.S. Cross-Channel Retail Forecast, 2012–2017*, October 29, 2013. http://www.forrester.com/US+CrossChannel+Retail+Forecast+2012+To+2017/fulltext/-/E-RES105461.

6. Intel Newsroom, Intel Labs Looks Inside the Future, June 25, 2013. http://newsroom.intel.com/community/intel_newsroom/blog/2013/06/25/intel-labs-looks-inside-the-future.

Chapter 5

1. http://bits.blogs.nytimes.com/2014/03/27/consortium-wants-standards-for-internet-of-things/?_php=true&_type=blogs&_r=0.

2. *RFID Journal*, May 5, 2014. http://www.rfidjournal.com/articles/view?11752/.

3. *Science Daily*. New Lab-on-a-Chip Device Overcomes Miniaturization Problems. April 30, 2014. http://www.sciencedaily.com/releases/2014/04/140430083143.htm.

4. *Journal of Chromatography B*, Volatile biomarkers from human melanoma cells http://www.sciencedirect.com/science/article/pii/S1570023213002730.

5. The University of Texas at Austin, Taste Chip Technology Description. http://research.cs.tamu.edu/prism/publications/ET_Broch.pdf.

6. Context Aware Computing for the Internet of Things: A Survey. *IEEE Communications Surveys and Tutorial (COMST)*, 2013. http://users.cecs.anu.edu.au/~charith/files/papers/J001.pdf.

Chapter 6

1. The Internet of Things Will Thrive by 2025. May 14, 2014. http://www.pewinternet.org/2014/05/14/internet-of-things/.

2. http://www.theatlantic.com/magazine/archive/2008/07/is-google-making-us-stupid/306868/.

3. Welcome to the Future Cloud—2025 in 100 Predictions.

4. The Gurus Speak | Pew Research Center's Internet & American Life Project: http://www.pewinternet.org/2014/05/14/the-gurus-speak-2/.

5. AP Impact: Recession, Tech Kill Middle-Class Jobs, http://bigstory.ap.org/article/ap-impact-recession-tech-kill-middle-class-jobs.

6. How Teens Do Research in the Digital World | Pew Research Center's Internet & American Life Project: http://www.pewinternet.org/2012/11/01/how-teens-do-research-in-the-digital-world/.

7. UCLA Newsroom. Is Technology Producing a Decline in Critical Thinking and Analysis? January 27, 2009. http://newsroom.ucla.edu/releases/is-technology-producing-a-decline-79127.

8. http://www.cafeny.ny.gov/bellevuestudy2013.pdf.

9. ISACA 2013 IT Risk/Reward Barometer. http://www.isaca.org/About-ISACA/Press-room/News-Releases/2013/Pages/ISACA-Survey-As-Internet-of-Things-Grows-Only-1-percent-of-Americans-Most-Trust-App-Makers-With-Personal-Data.aspx.

10. Big Data: Seizing Opportunities, Preserving Values. Executive Office of the President, May 2014. http://www.whitehouse.gov/sites/default/files/docs/big_data_privacy_report_may_1_2014.pdf.

11. How Companies Learn Your Secrets, *New York Times Magazine*, February 16, 2012. http://www.nytimes.com/2012/02/19/magazine/shopping-habits.html?pagewanted=1&_r=2&hp&.

12. FAA, Unmanned Aircraft Systems, FAA Aerospace Forecast Fiscal Years 2012–2032. http://www.faa.gov/about/office_org/headquarters_offices/apl/aviation_forecasts/aerospace_forecasts/2012-2032/media/Unmanned%20Aircraft%20Systems.pdf.

13. Robobees, Harvard University. http://robobees.seas.harvard.edu.

14. BAE Systems website. http://www.baesystems.com/magazine/BAES_026742/now-and-into-the-future.

15. Marc Goodman, A Vision of Crime in the Future. https://www.youtube.com/watch?v=-E97Kgi0sR4#t=53.

Chapter 7

1. Department of Transportation, Federal Aviation Administration. Interpretation of the Special Rule for Model Aircraft. Docket No. FAA-2014–0396 . http://www.faa.gov/about/initiatives/uas/media/model_aircraft_spec_rule.pdf.

2. McKinsey and Company, Insights and Publications, Disruptive technologies: Advances that will transform life, business, and the global economy May 2013. http://www.mckinsey.com/insights/business_technology/disruptive_technologies.

3. "Google's Eye in the Sky," by Will Oremus, *Slate*, June 13, 2014. http://www.slate.com/articles/technology/technology/2014/06/google_skybox_titan_aerospace_acquisitions_why_it_needs_satellites_and_drones.html.

4. Pew Research Internet Project, Digital Life in 2025. http://www.pewinternet.org/2014/03/11/digital-life-in-2025/.

FURTHER READINGS

Abbate, Janet. *Inventing the Internet*. MIT Press, 2000.

Armstrong, Stuart. *Smarter Than Us: The Rise of Machine Intelligence*. Machine Intelligence Research Institute, 2014.

Bardini, Thierry. *Bootstrapping: Douglas Engelbart, Coevolution, and the Origins of Personal Computing*. Stanford University Press, 2000.

Bauerlein, Mark. *The Digital Divide: Arguments for and against Facebook, Google, Texting, and the Age of Social Networking*. Tarcher, 2011.

Berners-Lee, Tim. *Weaving the Web: The Original Design and Ultimate Destiny of the World Wide Web*. HarperBusiness, 2000.

Beyer, Kurt. *Grace Hopper and the Invention of the Information Age*. MIT Press, 2009.

Bostom, Nick, *Superintelligence: Paths, Dangers, Strategies*. Oxford University Press, 2014.

Brynjolfsson, Erik, and Andres McAfee. *Race against the Machine: How the Digital Revolution Is Accelerating Innovation, Driving Productivity, and Irreversibly Transforming Employment and the Economy*. Digital Frontier Press, 2011.

Brynjolfsson, Erik, and Andrew McAfee. *The Second Machine Age: Work, Progress, and Prosperity in a Time of Brilliant Technologies*. Norton, 2014.

Carr, Nicholas. *The Big Switch: Rewiring the World, from Edison to Google*. W. Norton, 2013.

Carr, Nicholas. *The Shallows: What the Internet Is Doing to Our Brains*. Norton, 2011.

Hong, Sunghook. *Wireless: From Marconi's Black-Box to the Audion*. MIT Press, 2010.

Johnson, Deborah G., and Jameson M. Wetmore. *Technology and Society: Building Our Sociotechnical Future*. MIT Press, 2008.

Kamal, Devi. *Mobile Computing (Second Edition)*. Oxford University Press, 2012.

Karvinen, Tero, Kimmo Karvinen, and Ville Valtokari. *Make: Sensors: A Hands-on Primer for Monitoring the Real World with Arduino and Raspberry Pi.* Maker Media, 2014.

Kavis, Michael J. *Architecting the Cloud: Design Decisions for Cloud Computing Service Models.* Wiley. 2014.

Kurzweil, Ray. *The Singularity Is Near.* Viking, 2005.

McEwen, Adrian, and Hakim Cassimally. *Designing the Internet of Things.* Wiley, 2013.

Newman, Mark. *Networks, an Introduction.* Oxford University Press, 2010.

Norman, Donald A. *The Design of Future Things.* Basic Books, 2007.

Pew Research Internet Project. "The Internet of Things Will Thrive by 2025." 2014. http://www.pewinternet.org/files/2014/05/PIP_Internet-of-things_0514142.pdf

Reich, Pauline C., and Eduardo Gelbstein. *Law, Policy and Technology: Cyberterrorism, Information Warfare and Internet Immobilization.* IGI Global, 2012.

Rifkin, Jeremy. *The Zero Marginal Cost Society: The Internet of Things, the Collaborative Commons, and the Eclipse of Capitalism.* Macmillan, 2014.

Turkle, Sherry. *Alone Together: Why We Expect More from Technology and Less from Each Other.* Basic Books, 2011.

INDEX